W9-AJQ-761

ENERGY PROJECTS FOR YOUNG SCIENTISTS

Revised Edition

Richard C. Adams

and

Robert Gardner

FRANKLIN WATTS

A Division of Scholastic Inc.

New York ▪ Toronto ▪ London ▪ Auckland ▪ Sydney ▪
Mexico City ▪ New Delhi ▪ Hong Kong
Danbury, Connecticut

Photographs © 2002: Corbis Images: 15 bottom right, 130 (James L. Amos), 85 (Duomo), 17 (Kevin Fleming), 35 (Richard Hamilton Smith), 36 (Reuters New-Media Inc.), 82 (Ron Watts), 15 top; International Stock Photo/Tom O'Brien: cover; Peter Arnold Inc./BIOS: 107; Photo Researchers, NY/Jim Steinberg: 53; Rigoberto Quinteros: 127; Visuals Unlimited/Ken Lucas: 15 bottom left.

NOTE TO READERS:
Throughout this book, most measurements are given only in metric units because that is the system of measure used by most professional scientists. Words in italics appear in the Glossary at the back of this book.

Library of Congress Cataloging-in-Publication Data

Adams, Richard C. (Richard Crittenden)
Energy projects for young scientists / by Richard Adams and Robert Gardner. – Rev. ed.
p. cm. – (Projects for young scientists)
Previous ed. was authored by Robert Gardner.
Includes bibliographical references and index.
Summary: Instructions for a variety of projects and experiments demonstrating basic concepts of energy, work, and power, including thermal, electrical, and solar energy, energy of motion and position, and energy conservation.
ISBN 0-531-11666-2 (lib. bdg.) 0-531-16380-6 (pbk.)
1. Power (Mechanics)—Experiments. 2. Experiments. 3. Science projects.] I. Gardner, Robert, 1929- II. Gardner, Robert, 1929-. Energy projects for young scientists. 1987. III. Titile. IV. Series.
TJ163.95 .G37 2002
621.042'078—dc21

2001003033

1 2 3 4 5 6 7 8 9 10 R 11 10 09 08 07 06 05 04 03 02

CONTENTS

CHAPTER 1

SCIENCE PROJECTS AND FAIRS

Energy is a vital topic today. Energy costs rose dramatically during the 1970s, and even today most homeowners are concerned about their energy bills and the cost of heating or cooling their homes. Unfortunately, because of politics, multinational corporations, national rivalries, and other controlling factors, the economics of energy have little to do with the availability of oil or other energy sources. An understanding of energy, its sources, and ways to reduce our usage or improve the efficiency with which we use it will become increasingly important as some of our fossil-fuel energy resources approach depletion in the years to come.

While doing the energy projects suggested in this book, you may come up with new questions that you can answer through experiments of your own design. Be sure to carry out these experiments after consulting with an adult to be sure that they are safe. You are developing the kind of curiosity that characterizes a good scientist.

The purpose of this book is to provide some ideas that will help you get started on a project related to energy. Some of the projects suggested here simply raise questions that will make you think. You will have to design experimental procedures that answer the questions and respond to any other questions that arise as your inquiry proceeds.

SAFETY FIRST

1. Do all of the experiments and projects in this book under the supervision of a science teacher or other knowledgeable adult, unless such a person has approved of you working by yourself.
2. Read the instructions carefully before proceeding with a project. If you have questions, check with your "supervisor" before going any further.
3. Maintain a serious attitude while conducting experiments. Fooling around can be dangerous to you and to others.
4. WEAR PROTECTIVE GOGGLES AT ALL TIMES. Wear a lab apron when working with chemicals.
5. Do not touch chemicals with your bare hands unless instructed to do so. Wash your hands after conducting experiments involving chemicals.
6. Do not taste dry chemicals or solutions. Do not eat food while conducting experiments.
7. Do not inhale fumes released during a chemical reac-

tion. Perform experiments involving any poisonous or irritating gases under a fume hood.

8. Keep flammable materials away from heat sources.
9. Keep your work area clean and organized. Turn off gas and electricity when they are not being used.
10. Have safety equipment such as fire extinguishers, fire blankets, safety showers, and first-aid equipment available while you are experimenting. Know where this equipment is located.
11. Clean up chemical spills immediately. If you spill anything on your skin or clothing, rinse it off immediately with plenty of water, and then report what happened to a responsible adult.
12. Be careful about touching glass that has recently been heated; hot glass looks the same as cool glass. Bathe skin burns in cool water and apply ice.
13. Avoid touching materials connected to high voltage. Do not touch any source of high voltage.
14. Never experiment with household electricity unless supervised by a knowledgeable adult.

Other projects contain detailed instructions about how to proceed, but even these usually require some thinking on your part. If you have not had a chance to do much science on your own, you should probably begin with a project that provides reasonably detailed instructions. Once you've had more experience with science and experimentation, you may want to try some of the more open-ended projects, or you may discover questions of your own that you will want to explore. Projects of a more advanced nature are preceded by an asterisk (*). Many of the experiments are followed by a section entitled "Doing More." These are extensions to the experiment and will lead you into further explorations of the same topic. These can make a more extensive, complete science project.

Many people think of scientists as a group of men and women who possess miraculous memories and an abundance of factual information. The truth is that scientists, like people in all professions, are often forgetful and have to check books and notes frequently for the facts related to their research. What good scientists have in common is an ability to ask the right questions and a way to try to find the answers.

The science presented in textbooks is usually framed in terms of currently accepted theories about the way nature behaves. A reader learns very little about the scientific method from most textbooks. The variations and complexities of scientific investigation make it impossible to describe the scientific method. The only way to understand scientific inquiry is to do it. That is why doing science projects, when you don't know the outcome of the experiment or the right answer, can help you gain a better understanding of the nature of scientific inquiry.

Percy Bridgman, an American physicist, once said that the scientific method is doing one's utmost with one's mind, no holds barred. A more scholarly definition was given by James B. Conant: "Science emerges from other . . . activities of man to the extent that new concepts arise from experiments and observations." These new concepts or theories, he went on to say, stimulate the development of new experiments that confirm, modify, or cause us to discard the concept.

Conant's definition makes it clear that scientific

inquiry is a dynamic human process in which theories and ideas change as new information accumulates. The word *observation* in Conant's definition refers to the gathering of factual data, not just watching experiments as they take place. Gathering information requires the careful use of the investigator's sense organs, for our senses can be fooled easily. The observations we make may be planned, as when we set out to investigate a question that interests us, or they may be accidental. In more cases than the writers of textbooks would have us believe, crucial observations were made by accident.

Boyle's discovery of the law of gas pressures that bears his name arose from observations he made while attempting to disprove a theory proposed by another scientist. Darwin came to the idea of natural selection while reading a book by Malthus. Oersted discovered the connection between electricity and magnetism while preparing a demonstration designed to show that there was no connection between these two phenomena.

In an attempt to explain a set of observations, scientists often develop a working hypothesis. Developing such a hypothesis requires a "leap of the imagination" in attempting to present an explanation of the observations and the relationships among them. From this hypothesis, which is a generalization designed to explain factual information obtained by one or more investigators, scientists deduce specific consequences that follow from the hypothesis. They, or someone else, then design experiments that they hope will confirm or disprove the hypothesis. Even if experimental results confirm the prediction deduced from the hypothesis, the hypothesis is not proven. It merely gives additional evidence that it may be true.

Hypotheses and theories are never proven in science;

they only gain increasing certitude as more and more supporting evidence is acquired. If an experiment shows the hypothesis to be wrong, it may be modified or discarded.

When a broad working hypothesis, one that explains a large number of observations, has been confirmed by a great variety of experimentation, the term *theory* is often applied. For example, much of Dalton's hypothesis about the atomic nature of matter is now referred to as the atomic theory because it has been confirmed by repeated experimentation. Of course, this experimentation has gone on for 200 years and the theory has been modified and is still being modified to this day. This is what we mean when we say that science is a dynamic process. No explanation of nature's behavior is written in stone. There is always the possibility that today's explanation will be modified or replaced tomorrow.

What scientists come to after experimental checks and counterchecks, arguments, discussions, and tests of alternative explanations is a model or theory that explains some aspect of our universe. And what every scientist realizes is that that model may change as new information is gathered.

Underlying all scientific inquiry is the faith that nature is uniform, that experimental results obtained in New York today can be confirmed by similar experiments conducted in London next year. Only through such faith can science arrive at relationships and explanations that have meaning to all. Experimental results must be repeatable. If your data does not agree with that of someone else who does the same experiment, scientists will not accept the conclusions you make because they are based on data that cannot be confirmed.

In many cases, where a variety of causes of a phenomena are suspected, a scientist will try first one thing and then another in a trial and error fashion, looking for the one that gives the desired effect. This was the method used by Thomas Edison in much of his research. Unlike many scientists, Edison was not an analytical person who thought deeply about problems. He succeeded through great effort and determination, which probably explains his statement, "Genius is one percent inspiration and ninety-nine percent perspiration."

The purpose of this book is to provide as many ideas for energy projects as possible within the confines of these pages. If you are interested in using some of these projects as material for a science fair, you should consult a book devoted to the preparations you must make before entering such a fair. For the most current information on contests, competitions, and science fairs, check the Internet. More information about these books and web sites can be found in Appendix 3—Resources.

UNITS

You will find abbreviations for units used in this book. A list of these abbreviations and the units they represent can be found in Table 1.

TABLE 1: Units and Their Abbreviations

Length
foot (ft.)
inch (in.)
mile (mi.)
millimeter (mm)
centimeter (cm)
meter (m)
kilometer (km)

Area
square foot (sq. ft.)
square inch (sq. in.)
square centimeter (cm^2)
square meter (m^2)

Volume
cubic foot (cu. ft.)
cubic inch (cu. in.)
ounce (oz.)
milliliter (ml)
cubic centimeter (cm^3)
cubic meter (m^3)
liter (l)

Force/weight/mass
newton (N)
pound (lb.)
kilogram (kg)
gram (g)
milligram (mg)

Power
watt (W)
kilowatt (kW)
horsepower (hp)

Energy
joule (J)
newton-meter (N-m)
British thermal unit (Btu)
calorie (cal)
Calorie (Cal)
kilowatt-hour (kWh)
foot-pound (ft.-lb.)

Electrical units and their equivalents
volt (V) = joule/coulomb (J/C)
ampere (A) = coulomb/second (C/s)
coulomb (C) = ampere-second (A-s)
ohm (Ω) = volt/ampere (V/A)

Temperature
degree Celsius (°C)
degree Fahrenheit (°F)

Temperature-time
degree-day (DD)

Time
hour (h)
minute (min)
second (s)

Others
change in (Δ)
angle (θ)
kinetic energy (K.E.)
potential energy (P.E.)
gravitational potential energy (GPE)

CHAPTER 2

ENERGY, WORK, POWER, AND EFFICIENCY

Many people use the words *energy*, *work*, and *power* as if they all mean the same thing. Some science textbooks even define *energy* as "the ability to do work," but such a definition is an oversimplification.

ENERGY AND WORK

Work (W) is defined as the product of the net force (F) exerted on a body (in a direction parallel to the body's motion) and the distance (d) through which that force acts. In mathematical terms

Work = (Force) (distance)
or $W = Fd$
or better, $W = F\cos\theta\, d$,

where θ is the angle between the direction of the force and the direction in which the body moves.

If you lift a weight, W, a certain height, h, the work you do is Wh. If you lift the weight twice as high, you do twice as much work. But if the weight is too heavy for you to lift, you do no work when attempting to lift it. Even though you exert a large force, the distance through which the force acts is zero; consequently, the product of force and distance is zero.

Energy is a more subtle notion. You can't see, feel, touch, or taste energy. In fact, we can't really define energy. It is an intangible, yet, it appears in many familiar but different forms that can be defined. *Kinetic energy*, the energy of motion, is defined as $1/2\, mv^2$, where m is the mass of the moving body and v its velocity. Gravitational potential energy is defined as the product of a body's weight and its height above some level surface. Similarly, thermal energy (heat), electrical energy, light, elastic potential energy, as well as nuclear and chemical energy can all be defined.

The quality common to all forms of energy is the ability of one form to be transformed into another. For example, the gravitational potential energy stored in a ball lifted above the ground can be changed into kinetic energy as it falls. The kinetic energy is then transformed into elastic potential energy (the energy stored in any springlike device) as the ball is compressed upon striking the ground. The elastic energy is converted back into kinetic and then potential energy as the ball bounces upward. However, succeeding bounce heights become progressively smaller until finally that ball is at rest.

Careful measurements will show that the thermal energy (the total kinetic energy of the molecules of a sample of matter) added to the ball and its surroundings is

In a missile, energy produced by the burning of fuel is converted into thrust. A windmill converts solar energy in the form of wind into rotational energy, which in turn produces electrical energy. Solar energy also propels a bicycle. How's that? Plants utilize sunlight to create their own energy. Animals—including humans—derive some or all of their energy from plants. Humans then use some of this energy when they pedal bicycles!

equal to the original energy stored in the ball as gravitational potential energy. The various forms of energy associated with the falling and bouncing ball have been transformed into thermal energy.

Similarly, chemical energy (the energy stored in molecules) can be converted into the thermal energy needed to heat your home. The same energy can be used to produce steam that can turn an electrical turbine or drive a piston in the cylinder of a steam engine. The moving piston or electrical energy can then be used to do work. In either case, the amount of work done is always less than the amount of chemical energy utilized. In every energy transfer some thermal energy "rubs off" because of friction in bearings, air, or the movement of some surface over another.

The energy transformed from one form to another can be measured by the amount of work done. For example, suppose a ball that weighs 1 lb. falls from a height of 1 ft. and bounces to a height of ½ ft. We know that half the gravitational potential energy has been lost because the work required to lift a ball 6 in. is only half that required to lift the ball 1 ft.

POWER

Power is the rate at which work is done. A powerful engine is one that can do a lot of work in a short period of time. In mathematical terms

$$\text{Power} = \text{work/time, or } P = W/t$$

James Watt was the first person to measure power in units called *horsepower*. He measured the rate at which horses could raise heavy weights from a deep well and found that they averaged about 550 foot-pounds (ft.-lb.) of work per

Racing cars are powerful! They, too, are solar powered, since the oil they burn comes from decomposed plants.

second. (A foot-pound is the work done by a force of 1 lb. acting through a distance of 1 ft.) This value is still used today. When we say an engine has 100 horsepower (hp), it means that the machine can do 55,000 foot-pounds of work per second. Another unit of power is the watt (*W*), which is commonly used to measure electrical power. One watt is 1 joule per second (a joule, or newton-meter, is the work done by a force of 1 N acting through a distance of 1 m); a kilowatt (kW) is 1,000 W or 1,000 joules per second (1,000 J/s).

PROJECT 1: HOW POWERFUL ARE YOU?

What You Need	
Flight of stairs	Stopwatch

Do you think you can work like a horse? You may be surprised to find that you or one of your friends actually can work that hard.

When you walk up a flight of stairs, you are doing work on yourself. If you run up the stairs, you do the same amount of work in a shorter period of time and therefore develop more power.

To determine how much power you can develop, have someone measure the time it takes you to run up a long flight of stairs. To calculate your power you will need to know the *vertical* height of the stairs, your weight, and the time it took you to ascend the stairs. The product of your weight and the height of the stairs is the work you did on yourself. Dividing that work by the time required to do it will give you your power. For example, if you weigh 120 lb. and run up a flight of stairs 12 ft. high in 4.0 s, your power would be calculated as shown below.

$$\text{Power} = \frac{\text{weight} \times \text{height of stairs}}{\text{time}}$$

$$= \frac{120 \text{ lb.} \times 12 \text{ ft.}}{4.0 \text{ s}} = 360 \text{ ft.-lb./s} = 0.66 \text{ hp}$$

Try it! Are you as powerful as a horse?

Doing More

- Use the same method to find how much horsepower other people can develop. What do you find? Is the power that a person can develop related to age? Sex? Height? Weight? If you measure the power that people can develop over a greater height, say five flights of stairs instead of one or two, it will affect an individual's power. Can you explain why? It will be much easier if you can record your data in a computer spreadsheet to make your calculations easier. Most spreadsheets have graphing capability, too, and can make very good-looking graphs that are suitable for your final project presentation. If two factors are related, such as height and power, you should get a straight-line graph or a smooth curve. A program such as Vernier Software's *Graphical Analysis* will not only graph your data as you enter it, but will give you a mathematical equation that fits your data. Science fair judges (and teachers) are very impressed by equations derived from raw data.

EFFICIENCY

As you read earlier, some thermal energy always "rubs off" when energy is transformed from one form to another. Consequently, the amount of energy transferred into a machine is always greater than the energy that the machine transfers out. The efficiency of any device is the ratio of the energy input to the energy output. Since we always get less energy out of a device than we put into it, its efficiency is always less than 1.0.

PROJECT 2: HOW EFFICIENT IS YOUR BICYCLE?

What you Need	
Multispeed bike	Tape
String	Pair of sensitive spring scales

Many bicycles are built so that the rider can change gears. This makes it easier to pedal up hills. But when you make a bike easier to pedal, you sacrifice speed. You have to exert less force, but the bike's wheel doesn't go as far per turn of the sprocket gear attached to the pedals.

To measure the efficiency of a bicycle under the best conditions possible, that is, when the wheels are off the road and free to turn, you will need the equipment above. Turn the bike upside down. Use string and tape to attach one spring scale to a pedal and the other to the back tire. Figure 1 shows you how to do this.

Start the bike in low gear—the one you use to pedal up steep hills. Pull on the pedal with a fairly strong force. This is the input force. At the same time have a partner measure the force needed to keep the wheel from turning. This tells you the output force, the force that is exerted on the wheel and also the force that the wheel could exert on the road. Next, measure the distance that the pedal moves when it rotates once. Then, slowly and carefully, measure the distance a point on the tire moves when the pedal makes one turn.

You now have enough information to calculate the work put into the bike, the work that can come out of the bike, and, hence, the efficiency of the bike under ideal conditions. How efficient is the bicycle? What do you

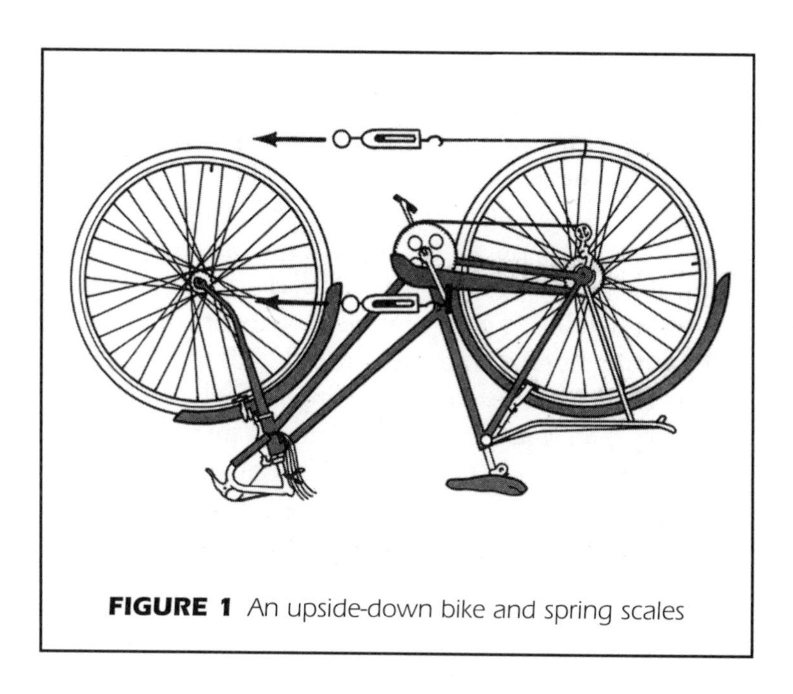

FIGURE 1 *An upside-down bike and spring scales*

find about the input and output forces, and the bike's efficiency, when you use different gears?

Doing More

- How could you measure the efficiency of the same bicycle under road conditions? One approach might be to let your full weight fall through the change in height that occurs when a pedal turns halfway around, from its highest to its lowest position. The distance that you fall while pushing on the pedal times your weight will equal the energy put into the bike. The kinetic energy of the bike and rider after the pedal reaches its lowest position can be used to measure the energy output.

- Find some other ways to measure the input and output of energies for the bicycle. One interesting correlation you can now do with modern test instruments is to take data on the heart rate of your rider. Heart-rate monitors, such as those available from Vernier Software, are usually of two types. One, a clip on the ear lobe, measures how light or dark the ear lobe is by how much blood goes through it. The other, attached to the chest, senses the electrical signals that make the heart beat. Because of the changing light conditions of a bicycle ride, the electrical sensor might be better for this project. Most of the Vernier probes can be connected to a computer or a Texas Instruments calculator, which makes portable data collection very practical.
- How efficient is the bicycle under road conditions? What is the effect of different gear ratios on its efficiency? Does the weight of the rider have an effect on the bike's efficiency? How about the air pressure in the tires? Look for smooth curves in a graph of these factors plotted against efficiency.

CHAPTER 3

HEAT AND THERMAL ENERGY

Matter is made up of atoms and molecules. These particles are in constant motion as they jostle one another about. The total kinetic energy of the molecules within a sample of matter is what we call thermal energy. The *average* kinetic energy of the molecules of a substance is proportional to the temperature of that substance. When the average kinetic energy doubles, the temperature doubles.

Heat is the thermal energy transferred from one body to another because of a difference in temperature. Heat "flows" from a warmer body to a cooler one. When the molecules of the warmer substance collide with those of the cooler substance, some kinetic energy is transferred through the collisions to the slower-moving molecules. If left in contact for some time, both samples of matter will come to the same temperature and, therefore, the same average kinetic energy.

The concept of heat is introduced early in this book because all other forms of energy can be converted into thermal energy. For example, the gravitational potential energy of a body can be transformed into thermal energy by attaching the body to a piece of plastic twine wrapped around a fixed metal cylinder. As the body falls, the plastic twine rubs against the metal and transfers heat to the metal, causing its temperature to rise. A famous scientist once demonstrated heat by letting a weight fall to the floor while a string attached to it turned a paddle wheel that heated some water by friction. (Look up Joule-Thompson heating some time.)

MEASURING HEAT

To determine how much heat is transferred from one body to another, we have to find a way to measure heat. The following projects will help you understand what factors are involved in taking such measurements.

To perform these projects you will need an immersion heater, the kind used to heat water for a single cup of tea. If you or your school do not own an immersion heater, you can buy one in a supermarket or department store. Be sure the heater is well insulated so that there is no danger of electric shock. When you disconnect such a heater, always pull the plug from the electric outlet. DO NOT PULL ON THE CORD THAT CONNECTS THE PLUG TO THE HEATER. NEVER INSERT THE PLUG INTO AN ELECTRICAL OUTLET UNLESS THE HEATER IS IN WATER OR ANOTHER LIQUID.

PROJECT 3: HEAT AND TEMPERATURE CHANGE

What You Need	
Immersion water heater	Graduated cylinder
Stopwatch	Thermometer or temperature probe
Foam cup	

To find out how the change in temperature of a substance is related to the amount of heat transferred to it, you can apply different amounts of heat to a fixed mass of water. If you were to change the mass of the water, you would not be able to tell whether the change in temperature was the effect of the heat or the mass. By keeping the mass fixed, you can be sure that you are testing the effect of heat on temperature change.

It seems reasonable to assume that the heat delivered by an immersion heater when it is plugged in for 30 s will always be the same, or "within the experimental error." By the same reasoning you can assume that an immersion heater plugged in for 60 s will deliver twice as much heat as one plugged in for 30 s. You can call the amount of heat that is transferred by an immersion heater in 30 s 1 glug of heat. Then the heat delivered by the immersion heater in 1 min will be 2 glugs, and so on.

Place 100 ml of cold water in a foam cup. Measure the temperature of the water with a thermometer calibrated in degrees Celsius. Heat the water with an immersion heater for 15 seconds. Stir the water with the heater as it warms the liquid. After 15 s pull the plug *but leave the heater in the cup* so that all the heat produced can be transferred to the water. Stir thoroughly and record the new temperature and the temperature change of the water.

Repeat the experiment several times. Use the same amount of water but increase the heat delivered from 0.5 glug to 1.0 glug, 1.5 glugs, and 2.0 glugs. Be sure to use a sample of cold water for each run because warm water loses heat to the surroundings faster than cold water.

How is the temperature change of the water related to the quantity of heat it receives?

Doing More

- To see how heat and mass are related, we will vary the mass of the water as we vary the heat in an effort to keep the temperature change constant. If heat is directly proportional to mass, doubling both the heat transferred and the mass of water heated should result in the same temperature change.

  Using your immersion heater, add one glug of heat to 100 g (100 ml) of cold water (the density of water is 1.0 g/ml so 100 ml of water has a mass of 100 g). Repeat the experiment using 200 g (200 ml) of cold water and 2 glugs of heat. In a large insulated cup (12 oz. or larger) try 3 glugs and 300 g and 4 glugs and 400 g. Are the temperature changes of the water close to being constant for all these runs? If they are, what does that tell you about the relationship between heat and mass?

- From what you have found experimentally about heat and mass and temperature change, what would you expect to find out about the temperature change if you kept the amount of heat transferred constant while increasing the mass of the water heated? To test your prediction, add 1 glug of heat to 100 g of cold water. What is the temperature change of the water? What is the temperature change when you add 1 glug of heat to 200 g of cold water? To 300 g of cold water? To 150 g of cold water? Is your prediction borne out by these experiments? Check it by graphing your results and looking for smooth curves.

What is the relationship between the mass of water and temperature change when the heat transferred is constant? What will a graph of mass versus temperature change look like when the heat transferred is kept constant? How about a graph of the inverse of mass (1/mass) versus temperature change? What equation can you use to express your findings?

- Your experiments have probably convinced you that heat is proportional to the temperature change of water and to the mass of the water heated. If something, in this case heat, is proportional to both of two variables, in this case temperature change and mass, then it (heat) is proportional to the product of these variables. For example, the area of a rectangle is proportional to both the length and the width of the rectangle.

  To see if heat is proportional to the product of mass and temperature change, make a chart like the one on the next page using the data you have collected in the three projects above. (A sample set of data, not necessarily the same as yours, is shown in the chart on the following page.)

  According to your data, is the heat transferred to the water proportional to the product of the water's mass and temperature change? Try a "linear" curve-fit to your data in *Graphical Analysis*. Does a more complex equation form give higher correlation?

Heat (glugs)	Mass of Water (g)	Temperature change (°C)	Temperature change × mass (g•°C)
1.0	150	13.0	1,950
1.0	100	20.0	2,000
1.0	200	10.0	2,000

UNITS OF HEAT

From the experiments above, you have seen that the amount of heat transferred to water is proportional to both the temperature change and the mass of the water. You could go on measuring heat in glugs if you wanted to, but few people would know what you were talking about. However, there are units of heat based on the proportionality between heat and the product of mass and temperature change that are commonly used in science and engineering.

One unit for measuring heat is the *calorie*. A calorie (cal) is defined as the amount of heat required to raise the temperature of 1 g of water through 1°C. The capital *Calorie* (Cal), used in nutrition and medicine, is defined as the amount of heat required to change the temperature of 1 kg water 1°C, so is actually a *kilocalorie* of the regular kind of calorie. Another common unit of heat used in engineering and the building industry is the British thermal unit (Btu). A Btu is the amount of heat required to change the temperature of 1 lb. of water by 1°F. The *joule* (J), a unit of heat used in science, particularly in physics, is the amount of heat needed to raise the temperature of 1 g water through 0.24°C. Physicists like joules because 1 J is also nearly equal to the energy needed to raise 100 g one meter in earth's gravity.

The units of heat discussed above are all based on the transfer of heat to water, but few substances can "absorb" as much heat per degree of temperature change as water. To take into account the differences in the quantity of heat that equal masses of different substances absorb, scientists introduced a notion called *specific heat*. The specific heat of a substance is the amount of heat needed to raise the temperature of 1 g of the substance through 1°C, or 1 lb of the substance through 1°F. Since it takes 1 cal to change the temperature of 1 g of water 1°C, the specific heat of water is 1 calorie per gram per degree Celsius.

PROJECT 4: SPECIFIC HEAT

What You Need	
Immersion heater Foam cup Graduated cylinder Stopwatch or watch with second hand Thermometer or temperature probe	Water, cooking oil, antifreeze Pieces of metal: aluminum, brass, iron, lead (for "Doing More" only) Balance (scale)

To find the specific heat of a liquid, begin by placing an immersion heater in 100 g of cold water in a foam cup. Measure the temperature change of the water after the heater is plugged into an electrical outlet for 30 s. (Be sure to stir the water and to leave the heater in the water after the plug is pulled so that the maximum amount of heat will be transferred.) How much heat, in calories, was transferred to the water? How much heat does the immersion heater "deliver" in 30 s? Look on the label of the immersion heater for a wattage rating and record that along with the time. Use a temperature probe connected to a computer to record the temperature change every second and see if you can relate the watt-seconds to the temperature increase with a graph. It will be helpful to know that 1 watt = 1 joule/sec and 4.18 joules = 1 calorie.

Repeat the experiment using 100 g of cold cooking oil. (Be sure the heater does not touch the foam cup. If it does, it may melt the material and give rise to a real mess when the oil leaks from the cup.) What *volume* of cooking oil will you need if you are to have a mass of 100 g? Remember, you have found the number of calories that the heater transfers in 30 s. In the same time an equal amount of heat will be transferred to the cooking oil. What was the temperature change of the cooking oil? What is the specific heat of cooking oil?

After rinsing the immersion heater thoroughly, determine the specific heat of antifreeze (ethylene glycol) using the same technique. What is the specific heat of antifreeze according to your data?

Doing More

- To determine the specific heat of a piece of metal such as aluminum, brass, iron, or lead, carefully place the metal in boiling water. Once the metal is at the temperature of the boiling water quickly (so as to lose as little heat as possible to the air), transfer the metal to some cold water in a foam cup. The heat transferred to the water obviously came from the hot metal.

Knowing the metal's mass and temperature change, calculate its specific heat. (Make sure you measure the temperature of the boiling water. At a higher elevation than sea level, or at a lower air pressure because of storms, water will boil at a lower temperature than 100°C.) For example, suppose a 20-g piece of aluminum at 100°C is added to 50 g of water at 20.0°C. If the water temperature rises to 26.5°C, then the metal transferred 325 cal of heat to the water. The heat transferred by the metal per gram per degree would be:

$$\frac{325\ \text{cal}}{(20\ \text{g}) \times (100^\circ - 26.5^\circ)} = \frac{325\ \text{cal}}{(20\ \text{g}) \times (73.5^\circ)}$$

$$= 0.22\ \text{cal/g-°C}$$

Thus, the specific heat of the aluminum is 0.22 cal/g-C.

Find some different metals and see if you can determine the specific heat of each one. You will have to adjust the mass of water and metal used to obtain reasonable and significant temperature changes.

- Collect some samples of different solid-metal elements. By measuring the density and specific heat of each one, you can identify each metal. Find a table of specific heats in the *Handbook of Chemistry and Physics* (U.S. Chemical Rubber Company), which your science teacher probably has, and see what other metals could be close to your values.

PROJECT 5: FLAME TEMPERATURES

What You Need	
Foam cup	Candle, Bunsen burner, alcohol burner, match
Graduated cylinder	Clothespin
Balance (scale)	Wire
Thermometer or temperature probe	Pyrometer (optional)
Pieces of metal (such as a steel washer)	

You can't measure the temperature of a flame by putting a thermometer into the flame; it will only break the thermometer. But suppose you add a small piece of metal, such as an iron or copper washer, whose mass and specific heat you know. You could hang the metal on a wire held by a clothespin, insert it into the flame, and heat the metal to the temperature of the flame. Then you could quickly drop it into some cold water.

How would this enable you to determine the flame's temperature? How would you know how long it took the metal to reach the temperature of the flame? Now check your results using a *pyrometer* hooked up to your computer (about $35 from Vernier Software). Are you close? Is there a correction factor you need to add or multiply by to make your "low-tech" method more accurate? The Vernier pyrometer is good for temperatures from –200°C to 1,400°C, but is only accurate to within 10 degrees. In doing your version of the experiment several times, how close are you to your average value each time? Ask a math teacher how to calculate standard deviation (or look in the Appendix of *Ideas for Science Projects* by Richard Adams and Robert Gardner, 1997).

Measure the temperature of a variety of flames—Bunsen burner, Bunsen burner with the air valve closed so it's a yellow flame, candle, alcohol burner, gas stove flame, and match. Are they all the same? **Be careful when using these sources of heat.**

SOURCES OF ENERGY FOR HOMES

All forms of energy can be transformed into thermal energy. In some of the following chapters, you will see that it is possible to change kinetic, potential, light, elastic, electric, and nuclear energy into thermal energy.

One source of energy that is becoming increasingly common as a means of heating water and homes is solar energy. The roofs of many homes support solar panels that face the sun and convert solar radiation into thermal energy that is used to heat water and/or air.

There are several sources of the kinetic energy in the giant turbines that turn electric generators in power plants and thereby produce electric energy. One source is the kinetic energy in water that has fallen from the tops of dams. A second is the steam produced by burning fossil fuels (coal, oil, or natural gas). In many of the power plants built in the 1950s through the 1980s, the thermal energy used to generate steam comes from the energy released when radioactive elements such as uranium and plutonium undergo nuclear fission. This process involves a loss in mass that gives rise to tremendous amounts of energy.

The high cost of fossil fuels—fuel, oil, coal, and natural gas—led many homeowners in areas in which wood was abundant to turn to wood-burning stoves and furnaces as means of heating their homes. But fossil fuels remain the chief source of heat for homes and for generating the steam that turns the turbines in electric power plants.

HEATING YOUR HOME AND SCHOOL

If you could turn up the thermostat in your home or school, raise the air temperature to a comfortable level, and then turn off the heat for the rest of the heating season, no one would complain about heating costs. Unfortunately, that's not the way it works.

When air in a building is heated, the air molecules acquire more kinetic energy. These fast-moving particles bump into the molecules on the surface of the walls of the building and transfer some of their kinetic energy to them. These molecules, in turn, transfer kinetic energy to molecules deeper in the walls. Eventually, kinetic energy is transferred by the molecules on the outer surface of the walls to less energetic molecules in the cold air outside the building. This transfer of energy from molecule to molecule through collisions is called *conduction*, and we say heat "flows" from warm objects to colder ones.

Infiltration of cold air into a building through cracks around windows, doors, and sills forces warm air out through other cracks. This type of heat transfer is called *convection*. There are also convection currents within a building as warm air rises over denser, cooler air; however, convection heat losses through ceilings, walls, and floors are generally small because the small air spaces within these structures break up convection currents.

Heat, like light, is transferred from the sun to earth by radiation. Since space is virtually a vacuum, there is no way the sun's heat could reach us through conduction or

Soldier's Grove was The United States of America's first solar village. After decades of repeated flooding, the residents of Soldier's Grove decided to build a new town center on higher ground. The new town was officially completed in 1983. During the rebuilding, the village passed ordinances stipulating that new buildings be built to specific thermal performance standards and obtain at least 50 percent of their heating needs with solar systems. Residents also passed a solar access ordinance to ensure that future buildings don't block the sun for existing structures.

convection. You have probably sensed radiant heat entering your body from a fireplace or from the sun when you stood in sunlight while wearing dark clothing. If you stand near a large window on a cold, cloudy day, you will

Solar heating and cooling system at a school in Honduras. Reflectors on the back slope of the collectors allow the collectors to absorb energy from both direct and reflected light.

probably feel cold as heat radiates from your body to the cold window.

The two major causes of heat loss from a building are conduction and infiltration. Conduction heat losses can be reduced by placing insulating materials in walls, ceilings, and floors. Sealing the cracks that allow cold air to enter a building will eliminate the drafts (infiltration) that drive warm air from a building and allow cold air to enter.

The projects that follow will help you determine the factors involved in conductive heat losses.

PROJECT 6: CONDUCTIVE HEAT LOSSES AND SURFACE AREA

What You Need	
Insulated container to hold water	Stopwatch or watch with a second hand
Plastic or metal molds to make different ice shapes	Balance (scale)
	Warm water

Your own experience has probably convinced you that surface area is related to the rate at which heat is lost to the surroundings. If you wake up feeling cold while a blizzard rages outside, you feel warmer when you curl up into a "ball." On the other hand (or season), you feel cooler on a hot, humid summer night if you stretch out, exposing as much of your body's surface area as possible to the air.

Establishing the relationship between conductive heat loss and surface area is not easy because other factors—insulation, temperature differences, and time—affect the rate of heat flow. As with any experiment, you must be careful to test only one factor at a time.

You can investigate the effect of surface area on conductive heat loss by making several pieces of ice that have the *same* volume but different surface areas. Pour 30 to 50 ml (but the same amount each time) of water into containers that have different shapes. You might use a cubic container and the top from a large plastic container to make a pancake-shaped ice "cake." You might also use cylindrical containers with very different diameters (don't use glass containers because they can break in the freezer). Because the ice cakes are made with the same volume of water, the same amount of heat will be required to melt each of them. Since 80 cal, 334 J, or 0.31 Btu are needed to melt 1 g of ice, you can easily figure out how much heat is required to melt whatever mass of ice is changed to liquid.

Before you add one of the ice cakes to the water, measure its dimensions (so that you can determine its surface area). Then quickly dry and weigh it. Drop the ice cake into the water. Stir the water constantly so that the surface of the ice will be in contact with the warm water, not with its own meltwater. After 10 s, remove the ice from the water. Quickly dry it and reweigh it. Repeat the procedure using the second ice cake, which has a different surface area. An alternate method of figuring out how much ice melted is to very carefully measure the volume of water after you have removed the ice. Any extra water has to have come from melted ice. Since 1 ml of water equals 1 g, you can determine the mass of melted ice.

How much mass of each of the ice cakes melted in 10 s? How much heat was transferred to one of the ice cakes per second? How much heat per second was transferred to the other piece of ice? What was the surface area of each piece of ice?

Compare the two ratios:

$$\frac{\text{Heat flow into ice cake \# 1 in 1 s}}{\text{Heat flow into ice cake \# 2 in 1 s}}$$

and

$$\frac{\text{Area of ice cake \# 1}}{\text{Area of ice cake \# 2}}$$

What do these ratios tell you about the relationship between heat flow and surface area? Now put all your data for heat flow and area into a spreadsheet, from smallest area to largest. Then graph your results (heat loss versus area) to see if you get a smooth curve or straight line. If it's a smooth curve, use *Graphical Analysis* to generate an equation. Do your results depend on straightline, square functions, or cube functions? Why can't you simply wait until all the ice melts and then calculate the heat flow per second? How will heating costs of a house with a large surface area compare with one with a small area if all other factors are the same? Why do soft drinks with crushed ice cool faster than ones with ice cubes? Using a paper and pencil drawing, "cut" a 10-cm ice cube into smaller and smaller cubes, calculating the total surface area for all your cubes with each step.

PROJECT 7: HEAT LOSS AND TEMPERATURE DIFFERENCE

What You Need	
2 foam cups Refrigerator	2 thermometers or temperature probes Stopwatch

From experience you know that the rate of heat flow from your body depends on the temperature of your surroundings. Step outside on a cold day and you'll lose heat so rapidly that you'll soon begin to shiver. On a hot summer day, you may feel uncomfortably warm even if you're clad only in light clothing because the rate of heat flow from your skin is so slow.

To find out how the difference in temperature between a warm body and a cool one affects the flow of heat, add 100 ml of hot water to each of two identical cups. The temperature of the water in both cups should be the same. Put one of the cups in a warm room; put the other one in a cold place such as a refrigerator. Record the temperature of the water in each container at 1-min intervals. Opening and closing the refrigerator door will introduce major variables here. If your science department has them, use temperature probes connected to a computer or calculator. The thin wires will fit easily in the refrigerator door gasket. In which place does heat flow more rapidly from the hot water?

To obtain a better understanding of how heat flow is related to temperature difference, plot graphs of temperature versus time for each cup of water on the same set of axes. Plot temperature on the vertical axis and time on the horizontal axis. In which environment does the water cool faster? From the slope (rise/run) of either graph, you can tell when the water is cooling fastest. How does the difference in temperature between the water and its environment affect heat flow? Use *Graphical Analysis* to make an equation for what's happening.

By drawing lines tangent to a carefully drawn graph of temperature versus time for hot water cooling in a cool environment, you can measure the rate of heat loss at various temperatures. The slope of the line will give you the change in temperature per unit of time. If it's not a simple straight line, *Graphical Analysis* can still give you a nice equation to fit your data. From the mass of water that is cooling and the temperature change, you can determine the rate at which heat is being lost at the temperature where the line is tangent to the curve (see Figure 2). Repeat this process at a variety of temperatures and note the temperature difference between the water and its cooler surroundings for each temperature. Then plot a graph of the rate of heat loss versus the difference in temperature between the water and its surroundings. How is the rate of heat loss related to the temperature difference between the warm water and the cooler air around it?

Doing More

- Design an experiment using identical ice cakes and large volumes of water at different temperatures to

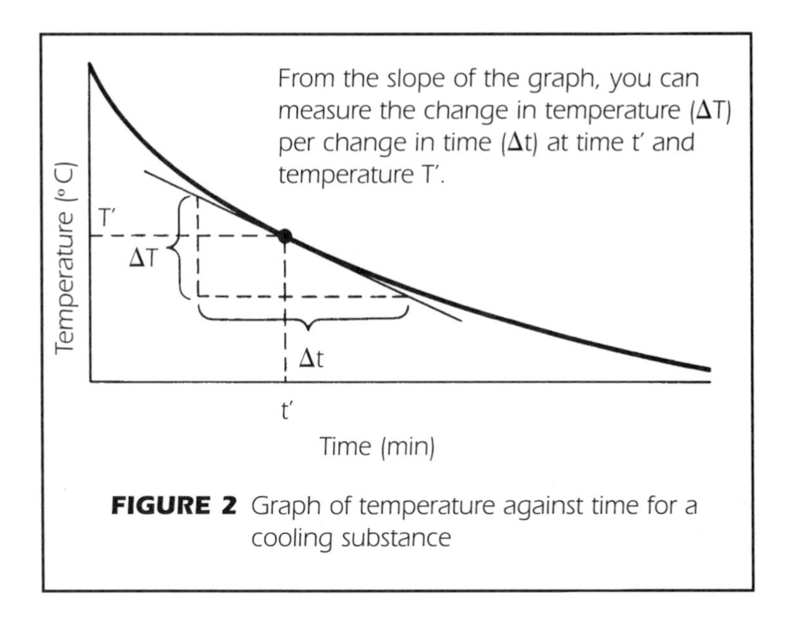

FIGURE 2 Graph of temperature against time for a cooling substance

see how the rate of melting is related to the temperature difference between the ice and the water in which it melts. Since the melting rate is determined by the rate at which heat flows into the ice, you are really finding how the rate of heat flow is related to the temperature difference between ice and its warmer surroundings. What do you find? Do your results agree with your findings in Project 7?

- Try using temperature probes to give you a continuous recording of the temperature change in both the ice and the water. Try it again with a smaller volume of water.

INSULATION AND R VALUES

In Projects 6 and 7 you found that heat losses are proportional to the time that heat flows. After all, if the temperature difference and the surface area of a warm body remain constant relative to some cooler surroundings, then in twice the time, twice as much heat will be transferred from the warm body. The only other factor involved in heat loss is the conductivity of the material that may lie between the warm body and its cooler surroundings. By "conductivity" we mean the rate at which heat flows through the material. Conductivity is measured by the amount of heat that flows through a unit area per unit time for each degree of temperature difference between the warm and cold side of the conducting material.

For example, suppose a warm air space surrounded by 10 sq. ft. of 3/4-in. wood sheathing loses 100 Btu of heat per hour when the temperature of the enclosed air is 10°F warmer than the air outside the wood. The conductivity of the wood is 100 Btu per hour per 10 sq. ft. per 10°F, or 1.0 Btu/h/ft.^{2}/°F. When fiberboard is used instead of sheathing, the heat losses drop to 50 Btu and so the conductivity, or U value, becomes 0.5 Btu/h/ft.^{2}/°F.

Homeowners and builders want insulating materials in the walls of their houses—substances that do not conduct heat well. Consequently, insulating materials are rated according to their ability to resist the flow of heat. A material's ability to resist the conduction of heat is measured in terms of R value, which is simply the inverse of conductivity. If we represent conductivity by the symbol U, then the R value of a material is defined as

$$R = 1/U$$

The conductivity of the fiberboard above was 0.5 Btu/h/ft.^{2}/°F. Its R value would be

$$R = 1/U = 1/0.5 = 2 \text{ ft.}^2 \times \text{h} \times \text{°F/Btu}.$$

R values are just U values turned upside down so that consumers can deal with numbers greater than 1. The larger the R value of a material, the better its insulating quality.

PROJECT 8: R VALUE

What You Need	
Large cardboard box 100-watt lightbulb and socket Tape, small blocks, ruler 2 thermometers or temperature probes	Various insulating materials: fiberglass, rock wool, cellulose, vermiculite, paper, foam, wood Stopwatch

To measure the R value of a cardboard box, put a 100-W lightbulb and socket in the center of a large cardboard box. Seal the box with tape and place it on several small blocks so that the entire surface of the box is exposed to the cooler air around it. Push the bulb of a thermometer or temperature probe through a small hole about halfway up the side of the box so you can measure the average temperature inside the box. Another thermometer or probe can be used to measure the temperature outside the box.

Plug in the lightbulb. When the temperature inside the box becomes constant, we can assume that heat losses from the box equal the rate at which heat is being generated by the bulb inside. We know the heat generated inside is 100 J/s because it's a 100-W bulb. Using this information, the area of the box, the temperature inside and outside the box, and a time interval of 1 s, determine the R value of the cardboard.

Now that you know the R value of a certain thickness of cardboard, you can determine the R value of other materials by comparing the rate at which they lose heat when compared to an equal area of cardboard. What is the R value of foam cups? What about the R values of other insulating materials such as fiberglass, rock wool, cellulose, vermiculite, paper, sheet foam, shredded foam, and wood? How is the R value related to the thickness of a material? Use a spreadsheet and *Graphical Analysis* to come up with an equation that represents your data.

How do the R values that you have measured compare with those found in books on home energy or at a lumber or building-materials store? Check the Appendix for sources of information.

DEGREE-DAYS

To predict the cost of heating a home for a year, heating engineers have developed the idea of degree-days (DD), which are based on degrees Fahrenheit. As long as the outside temperature is higher than 65°F, a building does not have to be heated. But when the outside temperature drops below 65°F, heating systems go to work. To determine the number of degree-days in one day, find the average temperature for the day in degrees Fahrenheit, subtract it from 65, and multiply by 1 day.

Suppose the average temperature for a certain day is 35°F. The degree-days for that day would be 30°F × 1 day, or, 30 DD. If the temperature for the next day averages 50°F, then the degree-days for that day would be 15 DD. The sum of the degree-days for the two days would be 45 DD.

Since the heat losses are proportional to the temperature difference between inside and outside air, a home will require twice as much heating fuel on a day when there are 30 DD as on a day when there are 15 DD. In a city such as Pittsburgh, Pennsylvania, where there are

5,987 DD in an average year, a homeowner will require about twice as much heating fuel for the same house as will a resident of Atlanta, Georgia, where there are only 2,961 DD in a year.

You can find the number of degree-days per heating season in your vicinity by reading the weather column in your local newspaper, visiting an Internet site such as http://www.weather.com, or by calling a fuel-supply company. If you multiply the number of degree-days by 24, you will have the number of degree-hours for your heating season. As a check on this system, you can easily mount a temperature probe on a long lead to a location away from a heated structure and record the temperature every hour or half hour. Use your readings to determine the "average temperature" each day and the degree-days in a heating season. Then compare your results with the values from the fuel-supply company for that time period.

You have seen that heat losses are proportional to time, surface area, temperature difference between inside and outside air, and the conductivity of the materials that enclose the warm air. Because a degree-day takes into account both the difference in temperature between inside and outside air and the time, and since conductivity, in Btu/h/ft.^{2}/°F, is the inverse of R values, we can calculate the heat losses from a building for the entire season using the following formula:

$$\text{Heat losses} = 1/R \times \text{surface area} \times 24\text{ h} \times \text{DD}$$

For example, a house with 2,000 sq. ft. of surface and an R value of 10 in a 4,000 DD area would lose:

$$1/10 \times 2{,}000 \times 24 \times 4{,}000 = 19{,}000{,}000 \text{ Btu of heat}$$

Of course, a house has walls, windows, ceilings, and often a basement, all with different R values, so it is usually necessary to consider conductive heat losses through all of these surfaces separately. By adding the heat losses through each surface, one can determine the total heat lost from the building in one heating season.

PROJECT 9: DETERMINING THE CONDUCTIVE HEAT LOSSES FROM YOUR HOME OR SCHOOL

What You Need	
Tape measure (50 ft. long if possible)	Thermometer

Measure the surface area of the walls, windows, and ceilings of your home or school. Using the degree-days for your town and the table of R values found in Appendix 1 or in similar tables that you can obtain from a lumberyard, determine the conductive heat losses from the building for one heating season. Another thing to consider is whether your home or school is kept heated to 65 degrees. In many European homes, it is customary to wear sweaters during the winter instead of heating to such high "summery" temperatures. How about *your* house?

Doing More

- Choose from the activities after Table 2.

INFILTRATION

Heat losses due to infiltration are not as easy to measure as conductive heat losses, but reasonable approximations can be made if you know the volume of the building, the number of air turnovers per hour, and the temperature difference between inside and outside air.

TABLE 2: Determining air turnovers per hour

	1 air turnover per hour if	2 air turnovers per hour if	3 air turnovers per hour if
Basement	Tight, no cracks. Sills, windows, doors all caulked and sealed.	Some cracks, loose windows and doors.	Big cracks, poorly sealed windows and doors.
Crawl space	Plywood floor, no trapdoor. Utility pipes entering building well sealed.	Tongue and groove floor, tight. Seals around trapdoor and utility pipes entering house.	Board floor. Loose fit on utility pipes entering house.
Windows	Tight-fitting, weather stripped, and storm windows.	Tight-fitting, weather stripped, but no storm windows.	Loose windows, no storm windows, no weather stripping.
Doors	Tight-fitting doors and storm door, weather stripping.	Loose storm doors. Poor fit on doors, no weather stripping.	No storm doors, doors fit poorly, no weather stripping.
Walls	Good caulking around all windows and doors. Building paper under siding.	Loose caulking, needs painting.	Cracks around doors and windows. No building paper under siding.

The volume of the air in a building multiplied by 0.018 Btu/ft.^{3}/°F, which is the heat required to raise the temperature of 1 cu. ft. of air through 1°F, gives the heat required to warm the air through a degree. If this figure is then multiplied by the number of air turnovers per hour, the number of hours in a day, and the number of degree-days in a year, you can find the heat required to warm the air that enters the building by infiltration. The formula below gives the heat losses associated with infiltration for one heating season.

$$\text{Heat} = \text{Volume} \times 0.018 \times N \times 24 \times \text{DD}$$

(N is the number of air turnovers per hour.)

The volume of air in a building is easy to find; all you need do is measure the volume of the air space within the structure. But how do you find N, the number of air turnovers per hour? This is not an easy task, but Table 2 will allow you to make a reasonable estimate of N. Simply average the four values obtained. You may decide to give ratings between 1 and 2 or 2 and 3; you might even give ratings of 0.5 if the seals are extremely tight. For example, if windows are tight and have double glazing, and there are storm windows as well, the windows would merit a 0.5 rating rather than 1.0. Extremely poor windows might receive a rating of 4. Triple glazing has become popular lately. Does this make a big difference? What information can you find out about this?

Doing More

- Determine the volume of air in your home or school. Then estimate the number of air turnovers per hour. Finally, determine the heat losses from your home or school that result from infiltration. For determining the turnovers per hour for a school room or the whole school, you might check with the building maintenance supervisor. There are laws in many locales that specify the minimum turnover required when the building is occupied by students. Of course, after hours, there may be considerable changes in the whole energy picture as lights go off, heaters shut down, blowers stop, and students are absent. Each person gives off as much heat as a 120-watt lightbulb. Having 30 students in a classroom can have a large impact on heating. Similarly, during the school day, every 45–60 minutes, all the doors inside and out may be opened and all the students in the school move from one place to another, moving masses of air with them. It makes a difference if they have to move to outside corridors to get from one part of campus to another or remain in the same building.

PROJECT 10: ANNUAL HEATING COSTS

Using heat losses due to conduction and infiltration that you found in Project 9 and the "Doing More" section, determine the total annual heat losses from your home or school.

Table 3 will enable you to convert heat in kilowatt-hours (kWh) of electricity, gallons of fuel oil, cords of wood, cubic feet of natural gas, and tons of coal to an equivalent number of Btus. Use the table to figure out how much energy is required to heat the building for a year. Then, after you find the cost of a unit of the type of fuel used to heat the building, determine how much it costs to heat the building for one year. Check with local oil, gas, and wood-stove dealers in your area to determine typical furnace efficiencies. Older furnaces use much more fuel than ones made after 1980. Many furnaces function for fifty years, so you'll find widely varying efficiencies of heat production.

PROJECT 11: ICE-CUBE KEEPING

What You Need	
Insulating materials	Ice
Construction cardboard	Stopwatch
2 thermometers or temperature probes	

Now that you are an expert on heat loss and insulation, put your knowledge to a test. Design some "ice-cube keepers," spaces in which ice cubes can be kept for long periods of time without melting even though they are in a room-temperature environment. See if you can predict which of a variety of keepers will be most effective. Use temperature probes to record the temperature of the interior over a long period of time. They may last more than 24 hours.

TABLE 3: Fuel efficiencies and conversion to Btus

Unit of heat	Equivalent No. of Btus	Approximate efficiency of fuel
1 kilowatt-hour	3,413	100%
1 ton of coal	26,200,000	60%
1 gallon # 2 fuel oil	138,000	70%
1 cubic foot of natural gas	1,031	70%
1 cord of wood	15,000,000	50%

CHAPTER 4

ELECTRICAL ENERGY

If you've ever turned on an electric stove or used an electric toaster in preparing breakfast, you know that electricity can be converted into heat. You've also probably seen electric motors lift weights, turn drill bits, spin lawn-mower blades, or do other forms of work, so you have good evidence that electricity is a form of energy. But where does electrical energy come from?

SOURCES OF ELECTRICITY

When an object such as a rubber rod is rubbed with a woolen cloth, both the rod and the cloth become electrified. If you take off a woolen sweater in a dark room during the winter, you see and hear the electric sparks produced when the positive and negative charges attract each other and combine. The charges that reside on surfaces that have been rubbed together constitute static electricity—electric charges at rest.

If an object that holds an excess of positive charge is connected by a wire to an object with an excess of negative charge, the negative charge (electrons) will move along the wire toward the positive charge because opposite charges attract one another. Moving charge constitutes an electric current, and electric current generally comes not from static charge, but from one of two sources—batteries or electric generators.

The chemical energy stored in batteries can be changed to electrical energy when the charge at one electrode of the battery flows to the opposite charge at the other electrode. In an electric generator, when a coil of wire is rotated in a magnetic field, electric charges are pushed along the wires. Through connections at the ends of the coil, charge can flow from one end of the coil through a circuit to the other end of the coil. This flow of charge is an electric current, just like the one that travels from one electrode of a battery to the other. The only difference is that with a generator, depending on how the ends of the coil are connected to brushes, the current can be made to flow periodically, first in one direction and then in the other (an alternating current), or continuously in the same direction (a direct current). With batteries, the current is always a direct current.

MEASURING ELECTRICITY

Electric meters, such as ammeters and voltmeters, are used to measure electricity. Ammeters measure electric current. If you examine an ammeter, you will see that the units on the meter are amperes (A). The ammeter needle remains steady when current flows because the meter measures the rate at which charge flows. It does not measure charge. If it did, the meter reading would start at zero and rise steadily as more charge flowed through the ammeter.

Generators at the McNary Dam powerhouse, at the Columbia River Gorge in Oregon.

PROJECT 12: AMMETERS IN SERIES AND IN PARALLEL

What You Need	
2 ammeters (0–1.0 A preferable) Voltmeter Insulated electrical wire (24-gauge or thicker), stainless steel electrodes	Several D-cell batteries Several flashlight bulbs and holders Stopwatch
For "Doing More" section, you will also need	
50-ml graduated cylinder Thermometer Beaker or pan of water, ringstand, clamp	25 g sodium carbonate monohydrate [$Na_2CO_3 \cdot H_2O$] Water

Ideally, instruments used to make measurements should have no effect on whatever they are measuring; however, a thermometer with a large mass might very well change the temperature of a small volume of water.

To see if an ammeter has a significant effect on the current it measures, set up the circuit shown in the top of Figure 3. If possible, use ammeters with a 0- to 1.0-A scale. If the current through the bulb is too small to measure accurately, add another D-cell to your battery, but don't let the circuit operate for very long; it could wear down the D-cells quite rapidly.

Find the current through the ammeter. Substitute a second, similar ammeter for the first one. How closely do the meter readings agree? Now connect the ammeters in series as shown in the bottom of Figure 3. Does the addition of

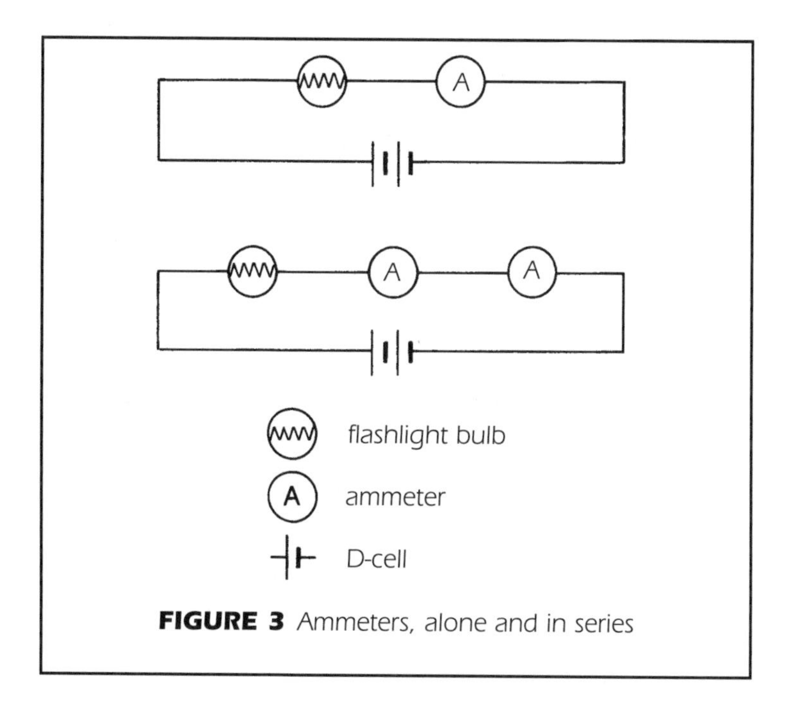

FIGURE 3 *Ammeters, alone and in series*

an ammeter to a circuit have a significant effect on the current? Substitute a voltmeter for one of the ammeters. Does a voltmeter have any effect on the current?

Using two bulbs and several D-cells, set up the parallel circuit shown in Figure 4. Use an ammeter to find the current through each bulb by placing the ammeter (or two identical ammeters, if you have them) at points x_1 and x_2. Now see if you can predict the current at points y_1 and y_2. How do you explain your results?

Doing More

- If an ammeter measures the rate of flow of electric current in amperes, then multiplying current by time (in seconds) should give us charge in units of ampere-seconds (A-s). But is there any way to be sure that the

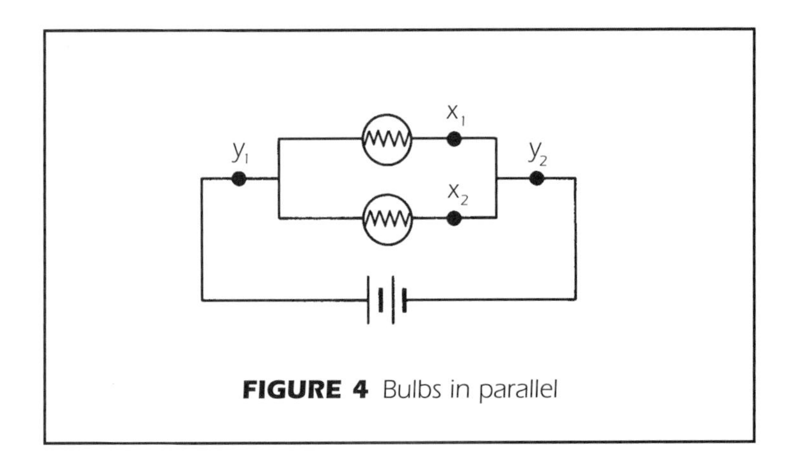

FIGURE 4 Bulbs in parallel

product of current and time is really a measurement of charge?

Certainty in science is hard to come by, but we become increasingly certain if an idea is confirmed in a number of different ways. During the electrolysis (decomposition by electricity) of water, hydrogen gas is released at the electrode attached to the negative pole of the battery, while oxygen is released at the other (positive) electrode. Since these gases are released only when the circuit is connected, we believe the flow of charge causes water to be decomposed into its elements. Therefore, it seems reasonable to believe that the mass (or volume under stable conditions) of these gases is directly related to the charge that has flowed through the circuit.

DO THIS EXPERIMENT UNDER SUPERVISION AND WEAR YOUR SAFETY GOGGLES.

To see if measuring charge as the product of current and time makes sense, set up the circuit shown in Figure 5. Note that the graduated cylinder is over the electrode connected to the negative pole of the battery. Thus, the gas collected in the cylinder will be hydrogen, which has twice the volume of the oxygen produced at the positive electrode. DO NOT COLLECT BOTH GASES. A MIXTURE OF HYDROGEN AND OXYGEN IS DANGEROUS BECAUSE A SMALL SPARK COULD CAUSE THE GASES TO EXPLODE, SENDING BROKEN GLASS IN ALL DIRECTIONS.

The solution in the beaker and cylinder is a 5% solution of sodium carbonate monohydrate [$Na_2CO_3 \cdot H_2O$]. To make the solution, add 25 g of sodium carbonate to 500 ml of water. You can find sodium carbonate in some stores sold as "washing soda" (not "baking soda").

Connect the circuit to the battery. Record the current, time, and volume of hydrogen collected at 60-s

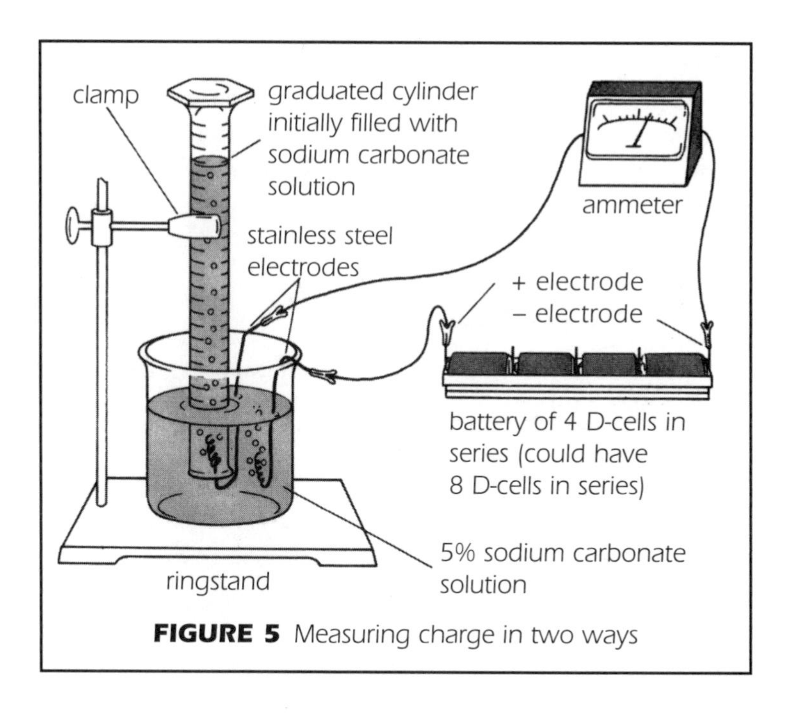

FIGURE 5 Measuring charge in two ways

intervals. Let the circuit run until about 30–40 ml of gas have accumulated before disconnecting it.

For each 60-s period, calculate the charge, in A-s, that flowed during that time. Then, by addition, make a running record of the total charge that flowed after 120 s, 180 s, 240 s, etc. The sample data table—table 4—below will help you set up your table. Use a spreadsheet to organize your data and make it easier to do the calculations.

Using your data, plot a graph of total charge, measured in ampere-seconds, on the vertical axis, versus total charge, measured in milliliters of hydrogen, on the horizontal axis. This is very easy to have your computer do if you recorded your data in a spreadsheet. What can you conclude from this graph? Do you have increasing evidence that charge can be measured by taking the product of current and time? Produce and analyze your graph with *Graphical Analysis*. What equation fits your data?

From the slope of the graph, determine the number of ampere-seconds equivalent to 1 ml of hydrogen. What volume of hydrogen is equivalent to 1 A-s? If your data is not a straight line, use the equation

TABLE 4: Sample Data

Time (s)	Volume of gas (ml)	Current (A)	Charge/60 s (A-s)	Total charge (A-s)
60	3.0	0.40	24	24
120	6.0	0.40	24	48
180	9.2	0.42	25	76
240	12.5	0.44	26	106

generated by *Graphical Analysis* to calculate your results.

If you know the barometric pressure and temperature of the surroundings (check http://www.weather.com for the air pressure for your city that day), determine the density of the hydrogen gas that you collected. However, you bubbled your hydrogen through water (primarily, although it also had dissolved sodium carbonate) and carried water vapor along with it. The gas in the tube was not just hydrogen, but hydrogen and water vapor mixed. Table 5 below gives the partial pressure of water at various temperatures. Just subtract the pressure given from the barometric pressure to get the pressure of the hydrogen alone in your cylinder. The density of hydrogen at 0°C and 1 atmosphere of pressure is

TABLE 5: Vapor Pressure of Water at various Temperatures (in torr)

Temp	Pressure	Temp	Pressure	Temp	Pressure
15	12.8	21	18.6	26	25.2
16	13.6	22	19.8	27	26.7
17	14.5	23	21.0	28	28.3
18	15.5	24	22.4	29	30.0
19	16.5	25	23.7	30	31.8
20	17.5				

760 torr is equal to 760 mm of mercury, 29.92 inches of mercury, 101.325 kilopascals, 1,000 millibars, 1 atmosphere, or 15 lbs. per square inch. Use a proportion to convert to the units of barometric pressure you need.

0.0899 g/l or 8.99×10^{-5} g/ml. Since your gas probably was not at 0°C and not at one atmosphere pressure, use the following formula to determine what the volume would have been under those conditions:

$$V_2 = \frac{V_1 \times (T_2) \times (P_1)}{(T_1) \times (P_2)}$$

V_1 is the volume you measured, T_1 is the temperature of the room, T_2 is zero degrees Celsius (273°K), P_1 is the pressure you measured (minus the correction factor due to water vapor), and P_2 is one atmosphere. You can use whatever units you want for the pressure and volume, but the temperature *must* be in degrees Kelvin. To get degrees on the Kelvin temperature scale, just add 273 to your Celsius measurements, or

$$K = °C + 273$$

Knowing what volume of dry hydrogen you would have had under standard conditions, what mass of hydrogen is produced by 1 A-s of charge?

PROJECT 13: WHAT DOES A VOLTMETER MEASURE?

What You Need	
Voltmeter, ammeter, DC variable-voltage power source	Foam cups
20-ohm resistor (20 Ω)	Insulated wire, alligator clips
Graduated cylinder	Stopwatch
Thermometer or temperature probe	

You've probably heard that a voltmeter measures voltage, or potential difference. But what does that mean? To find out, connect a voltmeter with a 0 to15-V scale in parallel with an electric heater and ammeter as shown in Figure 6.

A 20-ohm resistor can serve as the heater. Place it in 50 g of cold water in a foam cup. The temperature of the water should be about 2–3°C below room temperature to compensate for heat loss when the water temperature rises above that of the room.

Record the temperature of the cool water in the insulated cup to the nearest 0.10°C, if possible. Set the variable voltage source so that a current of approximately 0.7 A flows through the heater immersed in water for a period of 2 min. Record the voltmeter and ammeter readings as the current flows. Temperature probes can record temperatures every minute to within 0.10°C easily. After the trial is completed, stir the water thoroughly and record its final temperature. Did the heating take place at a constant rate? You can determine that if you were able to use temperature probes. Use *Graphical Analysis* to see if the temperature change was a *linear* change (a straight line).

What was the change in temperature of the water? How much heat, in calories, was produced by the electricity?

In succeeding trials, starting with fresh cool water each time, let a current of about 0.6 A flow for 2.5 min, then 0.5 A for 3.5 min, and finally, 0.4 A for 5 min. Record the voltage, current, time, and temperature change of the water for each trial.

When you have collected all the data, calculate the heat delivered, in calories, and the charge, in A-s, for each run. Then plot a graph of *heat per charge* on the vertical axis versus the voltmeter reading, in volts. What does the graph tell you? What does a voltmeter measure?

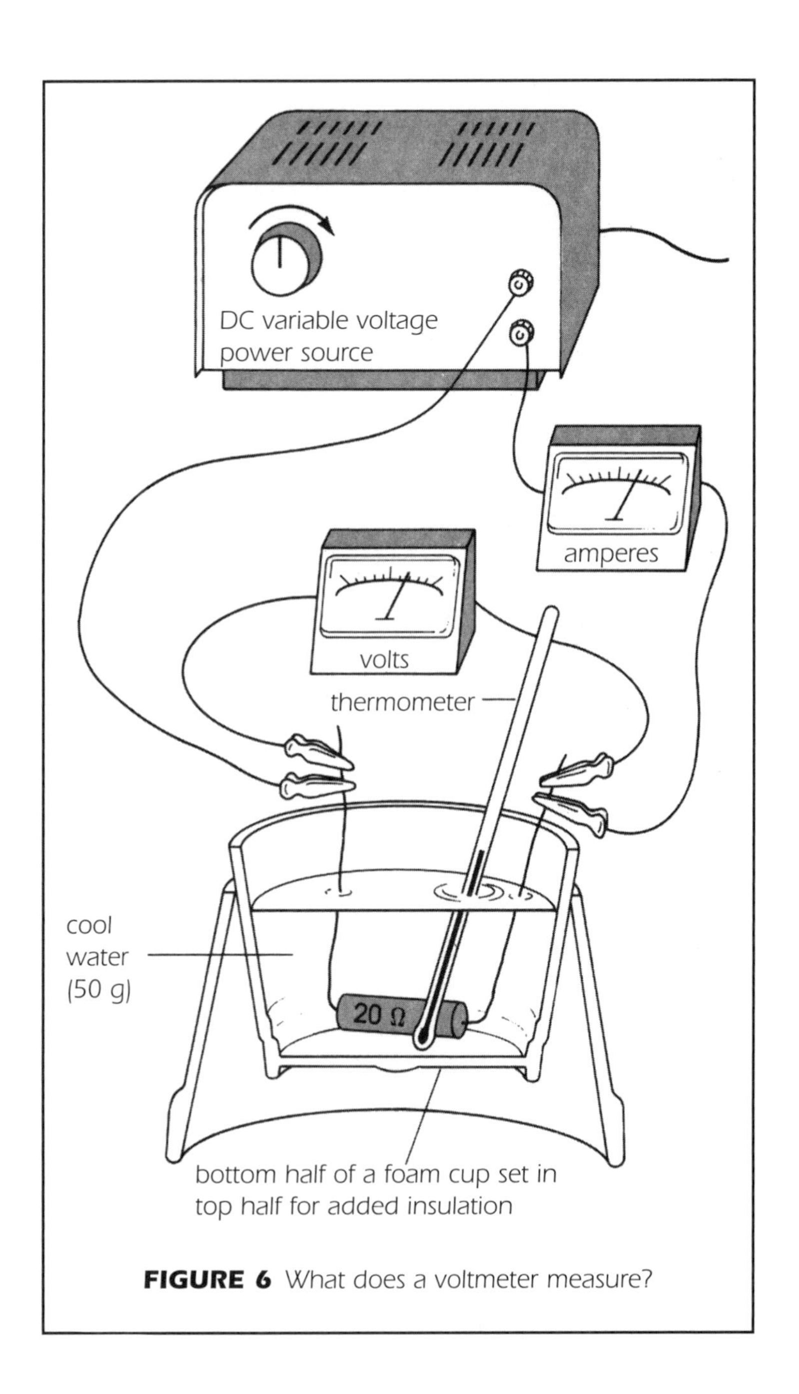

FIGURE 6 What does a voltmeter measure?

Write an equation relating heat, current, time, and voltage. *Graphical Analysis* can give you a start. Input your data, have the program graph it, and then try different families of equations until you get a close match.

The product of voltage $\times$ current $\times$ time is called *electric work* and is measured in joules. If you measure heat in joules, write the equation relating heat, current, time, and voltage. How many joules are equal to 1 cal?

Notice that you may now write 1 V = 1 J/A-s.

Charge is also measured in another unit called a coulomb (C). One coulomb is the same amount of charge as 1 A-s; therefore, you can also write

$$1 \text{ V} = 1 \text{ J/C}$$

PROJECT 14: METERS IN YOUR HOME AND SCHOOL

Most of the electricity used in homes and business is purchased from utility companies. To determine how much to charge each month, the utility company puts a meter in the line leading to each building for which a bill is rendered.

If you can find the place where the electric power line enters your house, apartment, or school, it will be connected to a meter. Current flowing through the meter causes the disk inside the meter to turn. Through a series of gears, the spinning disk turns the meter dials, and this gives a reading of the electrical energy entering the building. The dial on the right records kilowatt-hours, the next dial indicates tens of kilowatt-hours, the next hundreds of kilowatt-hours, and so on.

As you probably know, a watt is a unit of power equal to 1 J/s. An ampere of current is a flow of 1 C/s. Hence, you may write:

$$1 \text{ C/s} \times 1 \text{ J/C} = 1 \text{ J/s or } 1 \text{ W}$$

A kilowatt is 1,000 J/s, and an hour is equal to 3,600 s; therefore, a kWh is: 1,000 J/s × 3,600 s = 3,600,000 J.

The power ratings on various appliances, when multiplied by the time they are used, will indicate how much electrical energy these devices require. For example, a 100-W lightbulb, when turned on for 1 h, will require 0.10 kW × 1.0 h = 0.10 kWh. If the power company charges you $0.10 per kilowatt-hour, operating the lightbulb for 1 h will cost 0.10 kWh × $0.10/KWh = $0.01, or one cent.

Make a survey of the electrical appliances in your home or school. How many hours or minutes per day does each one operate? How much energy does each appliance require over a year's time? Somewhere on your family's or school's electric bill you will find the cost of a kilowatt-hour. Use this figure to determine the cost of operating each appliance for one year. If you add in all the appliances, including the water heater and furnace (if electric), how close do you come to your family's monthly bill?

Check out Table 6 in Chapter 7 to see the usage of a typical family.

THE DIRECTION OF CURRENT FLOW

That fact that electric current disappears when a circuit is broken at any point is good evidence to support the idea that charge flows all the way around a circuit. But which way does charge flow? Does positive charge flow from the positive to the negative end of a battery? Or does negative charge flow in the opposite direction? Or does charge of both signs flow in both directions?

Physicists defined the direction of electric current to be the direction that positive charge would flow in a circuit long before the actual direction was determined. That definition still holds, and, as you will see, that definition is sometimes valid.

ELECTROLYSIS, ATOM COUNTING, AND THE CHARGE PER ATOM

Historically, scientists were able to determine the relative charge associated with atoms of different elements before the actual masses of the atoms were known. Today, the actual masses of all the atoms have been determined. You can easily find the mass of an atom of any element. Simply divide the atomic mass of the element, in atomic mass units (amu), by 6.02×10^{23}. The mass of the atom will be in grams.

Knowing the mass of a single atom, you can calculate how many atoms are plated out or dissolved during electrolysis. If you also know the total charge involved in bringing about these mass changes, you will be able to determine the charge associated with each atom as well.

PROJECT 15: ELECTROLYSIS AND THE CHARGE PER ATOM

What You Need	
2 zinc electrodes (2 by 4 inches of sheet zinc)	Several D-cell batteries or a low-voltage power source
287 g zinc sulfate heptahydrate [$ZnSO_4 \cdot 7H_2O$]	Ammeter (0–1.0 A)
Insulated wires, alligator clips	Balance (scale)
1-liter container	Ringstand, insulated clamps
Graduated cylinder (1 liter)	Paper towels
Steel wool	Heat lamp
	Stopwatch

For the "Doing More" section, you will also need	
Lead electrodes of the same size as the zinc ones	75 ml of 6 M sulfuric acid (CAUTION!)
Copper electrodes	1 g unflavored gelatin
165 g lead nitrate [$Pb(NO_3)_2$]	3 liters distilled water
200 g hydrated copper sulfate [$CuSO_4 \bullet 5H_2O$]	2 more 1-liter containers
	Safety goggles, lab apron, rubber gloves

If you did the "Doing More" experiment from Project 12, you may have found that 1.04×10^{-5} g of hydrogen are released for every ampere-second of charge that flows during the electrolysis of water. Since the mass of a hydrogen atom is 1.66×10^{-24} g, 1 A-s of charge is associated with 6.27×10^{18} atoms of hydrogen. This means that the charge required to release one atom of hydrogen during electrolysis is

$$1.0 \text{ A-s} / 6.27 \times 10^{18} \text{ atoms} = 1.60 \times 10^{-19} \text{ A-s/atom.}$$

Because one atom of oxygen is released for every two atoms of hydrogen during the electrolysis of water, the charge required to release one atom of oxygen must be just twice that for hydrogen, or 3.20×10^{-19} A-s.

To find the charge associated with one atom of zinc, place a pair of zinc electrodes in a 1.0 M solution of zinc sulfate ($ZnSO_4$) as shown in Figure 7.

You can make the solution by dissolving 287 g of $ZnSO_4 \bullet 7H_2O$ in distilled water and then diluting to 1 liter with water. The electrodes can be made by cutting 2 $\times$ 4 in. (5 cm $\times$ 10 cm) pieces from a sheet of zinc.

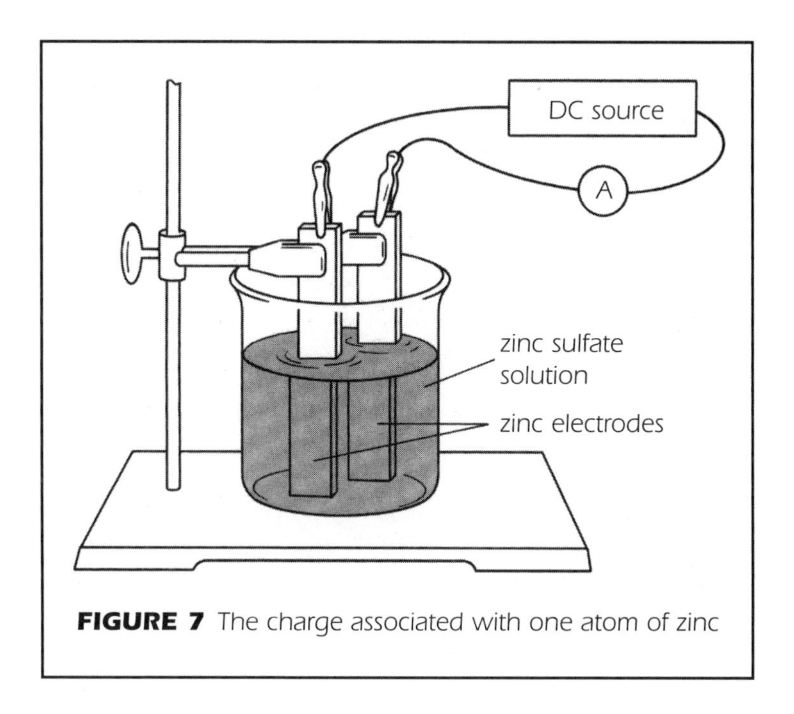

FIGURE 7 The charge associated with one atom of zinc

Clean the electrodes with steel wool or a scouring pad and then weigh each electrode to the nearest 0.01 g. Using D-cell batteries or a low-voltage DC power source, let a current of about 1.0 A run for approximately 30 min. Record the current every minute.

After disconnecting the circuit, gently rinse and dry the electrodes by dipping them in a beaker of water and then patting them dry with a paper towel. To ensure thorough drying, place the electrodes under a heat lamp for several minutes. Then weigh the electrodes. After several minutes, weigh them again to be sure that they were thoroughly dry.

Find the average current and multiply it by the total time in seconds. This will give you the total charge that flowed through the circuit.

Which electrode gained mass? Which electrode lost mass? How do the mass changes for the electrodes compare? Can you explain what is happening during the electrolysis of zinc sulfate? How many atoms of zinc came out of, or went into, solution at each electrode? What charge, in ampere-seconds (or coulombs), is associated with one atom of zinc?

Try to predict the mass changes that will occur at each electrode if you operate the cell for different times at about the same current.

Doing More

- You can determine the charge associated with atoms of other elements in a similar manner.

  For lead, use electrodes of the same size cut from a sheet of lead. The solution can be prepared by dissolving 165 g of lead nitrate [$Pb(NO_3)_2$] and 1 g of gelatin in enough warm, distilled water to make a liter of solution. (Somehow, the gelatin makes the lead stick better to the negative electrode.) To avoid the growth of "trees" during electrolysis, bend the electrodes into a convex shape by gently pressing them against the side of a beaker. Then set them in the solution, after weighing, with their edges as far apart as possible (convex surfaces facing each other).

  For copper, use similar-size electrodes cut from 20-mesh copper gauze. A solution of copper sulfate can be prepared by dissolving 200 g of hydrated copper sulfate [$CuSO_4 \bullet 5H_2O$] in enough water to make about 0.9 liter of solution. Slowly add 75 ml of 6 *M* sulfuric acid (*careful!*), with lots of stirring, and dilute with distilled water to 1 liter.

  THIS EXPERIMENT INVOLVES THE USE OF POTENTIALLY DANGEROUS CHEMICALS. BE SURE TO WEAR SAFETY GOGGLES

AND A LAB APRON AND TO WORK UNDER THE SUPERVISION OF A KNOWLEDGEABLE ADULT. KEEP A FAUCET RUNNING WHILE YOU DO THIS TO IMMEDIATELY RINSE OFF ANY SPILLS TO YOUR HANDS AND KEEP A BOX OF BAKING SODA HANDY TO NEUTRALIZE ANY SPILLS. KEEP POURING BAKING SODA ONTO AN ACID SPILL UNTIL IT STOPS FIZZING.

ELECTRIC CELLS

One of the simplest and oldest electric cells is the Daniell cell. It was invented in 1836 by John F. Daniell. It was the first cell that could be depended upon to give a steady current for long periods, and was used extensively in the early days of telegraphy to send electric signals along wires.

PROJECT 16: BUILDING AND USING A DANIELL CELL

What You Need	
Foam cup	Voltmeter or voltage probe or digital voltmeter
Parchment paper, glue	Ammeter
Copper and zinc electrodes	Safety goggles
Insulated wires, alligator clips	Stopwatch
Solutions of copper sulfate and zinc sulfate from Project 15	

Figure 8 shows you how to build a simple Daniell cell using a foam cup, parchment paper, and glue. Once the container is built and the glue has dried, electrodes of copper and zinc may be placed on opposite sides of the parchment barrier and secured with clamps. The bottoms of both electrodes should be just above the bottoms of the cups. Solutions of zinc sulfate and copper sulfate, prepared in Project 15, will be poured into the compartments containing the zinc and copper electrodes to provide electrolytes for the cell. Minimize mixing of solutions due to diffusion through the barrier by not adding the solutions to the cell until you are ready to measure current, temperature, and time.

WEAR SAFETY GOGGLES AND WORK UNDER SUPERVISION.

Pour copper sulfate solution into the side of the cup that contains the copper electrode. Pour zinc sulfate into the other side. Measure the temperature of each solution, then connect a voltmeter to the electrodes. What is the potential difference (voltage) across this cell? Connect a sensitive ammeter to the electrodes so it is in parallel with the voltmeter. Move the electrodes closer together, if necessary, to obtain a current of at least 0.5 A.

What happens to the voltmeter reading when current flows? What happens to the temperature of the solutions as time passes? Record the current every minute for at least 20 min. To collect your data more easily and accurately, use a voltage probe instrument connected to a computer or calculator.

Disconnect the circuit, rinse and dry both electrodes as described in Project 24, and determine the change in mass of each. Which electrode gained mass? Which one lost mass? Determine the total charge that flowed in the circuit.

With a razor blade, carefully cut a slit along the dotted lines as shown in the drawing.

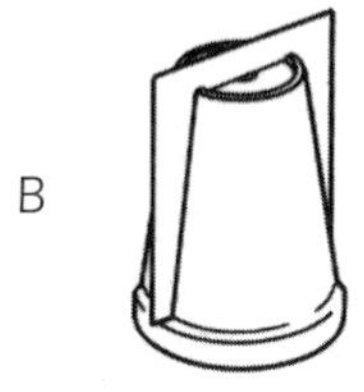

Slide a piece of parchment paper into the slit as shown.

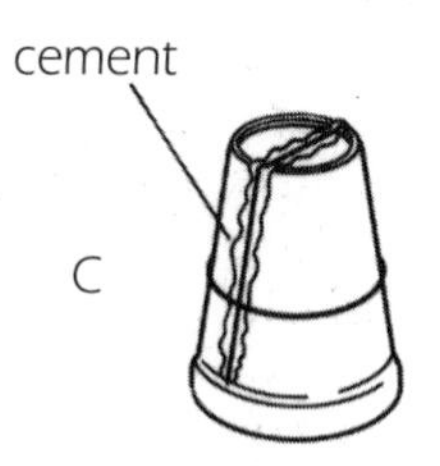

Carefully trim the parchment paper to fit the cup. Seal with Duco cement and use a rubber band to hold cup firmly against paper until cement dries.

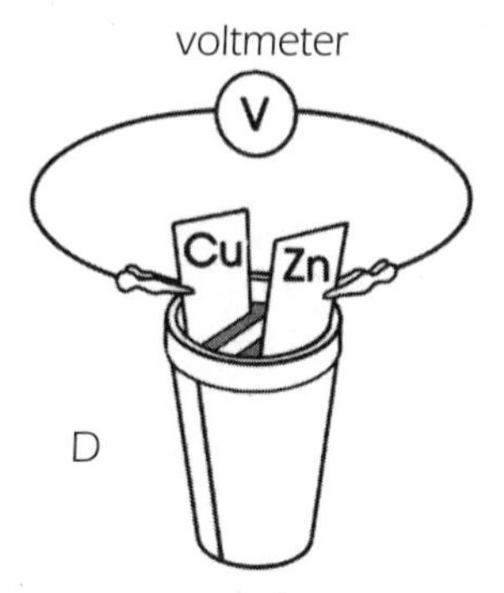

When dry, a zinc electrode and $ZnSO_4$ solution can be placed in one side and a copper electrode and $CuSO_4$ solution in the other.

FIGURE 8 Building a Daniell cell

From your measurements of charge and mass, calculate the number of atoms of each element that plated out or dissolved. Determine the charge associated with one atom of zinc and one atom of copper. How do your results of charge per atom compare with those found in Project 15?

From what you have seen and measured in this experiment, try to devise a model to explain the flow of charge in this circuit. Remember, only electrons can move along wires, and, in the solutions, both copper and zinc ions carry positive charge. Sulfate ions, which are also free to move in solution, provide the negative charge needed to balance the charges on the positive ions.

Doing More

- Repeat the experiment, but this time watch what happens to the voltmeter reading as the current through the circuit changes. You may be able to get better accuracy using a digital voltmeter. To change the current, simply change the distance between the electrodes. What happens to the voltage as the current increases? Use a voltage probe connected to a computer to record the data as you move the electrodes measured distances. Use *Graphical Analysis* to get a mathematical relationship between voltage and distance.

  From Project 13 you learned that a voltmeter measures energy per charge. A voltmeter reading of 10 V means 10 J/C are being delivered to the part of the circuit between the leads of the voltmeter. Can you explain why the voltage decreases as the current through the cell increases? If not, consider this question: Does the electric cell have an electrical resistance? The electrical resistance of a circuit element is defined as the ratio of the voltmeter reading across the element to the current flowing through it:

$$R = V/I$$

where R is the resistance, V the voltage, and I the current. In the Daniell cell you have built, there is practically no resistance through the ammeter and wires that connect it to the cell. Hence, when the ammeter is connected to the electrodes, the voltmeter, which measures the product of IR in the circuit outside the cell, will read close to zero. So where is the energy per charge that the voltmeter indicated was present before the ammeter was connected? Leave a Daniell cell for several hours. Is there any movement of ions through the parchment barrier? How can you tell?

- To see why a barrier is used in such a cell, try this. Place a small piece of zinc in a small amount of copper sulfate solution. What happens?

  Next, use wire leads to connect two small zinc electrodes to a sensitive ammeter. Dip the electrodes first into zinc sulfate and then into copper sulfate solutions. If you observe any current in either solution, try jiggling first one electrode and then the other. What do you observe?

  Repeat the process using two small copper electrodes. Then repeat again, but this time use one small copper and one small zinc electrode. Finally, try the experiment using zinc in a solution of copper sulfate but in separate containers. Do you observe any current? Do you observe any current if you connect the two solutions with a piece of paper towel soaked in one of the solutions?

  On the basis of your experiments and observations, why do you think a barrier is used in a Daniell cell?

- To store the electric energy generated from solar cells requires batteries that can be recharged thousands of times. If electric cars are to be a practical means of transportation, it will be necessary to develop a similar type of battery. Can you invent such a battery?

PROJECT 17: MAKING AN ELECTRIC MOTOR

What You Need	
6-volt battery or 4 D-cell batteries	Small magnet
Various pieces of copper metal and insulated and uninsulated wires	Toy wagon or other small toy vehicle

There are three basic parts to a motor: a stator, an armature, and brushes. Design and build a DC electric motor that will run on a 6-V battery. DO NOT ATTEMPT TO USE ELECTRICITY FROM AN ELECTRIC OUTLET IN YOUR HOUSE. Look for diagrams of small motors in physics books.

Doing More

- Install your electric motor and battery in a small toy vehicle and make it move under its own power.

PROJECT 18: MAGNETISM FROM ELECTRICITY

What You Need	
Magnetic compass 2 D-cell batteries (You can use more for the "Doing More" section.)	Long, straight copper wire Clay for support of wires
For the second "Doing More" section, you will also need	
Galvanometer	Coil of wire

Before 1819, most scientists thought that electricity and magnetism were unrelated, but in that year Hans Christian Oersted, a young Danish physicist, accidentally discovered that a magnetic field surrounds an electric current. He was trying to show his students that electricity and magnetism were not related when his demonstration showed the opposite. A surprising number of scientific breakthroughs have happened because of experiments that went "wrong."

To see Oersted's "failed" experiment for yourself, place a magnetic compass above and touching a long, straight wire. The wire should be parallel to the compass needle. Carefully touch the ends of the wire to the poles of a battery made from two D-cells in series. (Don't let current flow for very long or the battery will wear out quickly.) What happens to the compass needle when charge flows along the wire? You have now seen what surprised Oersted and his students more than 180 years ago.

Now place the wire above the compass needle and let

current flow through the wire again. Which way does the needle turn this time? What happens to the direction of the magnetic field if you reverse the direction of the current by turning the battery around?

Try moving the compass needle farther from the wire. How far away from the wire can you detect a magnetic effect when current is flowing? If you use a stronger battery, up to eight D-cells in series, can you detect any change in the strength of the magnetic field when a current flows? Can you offer a hypothesis to explain your observations? How can you test your hypothesis?

Doing More

- Design experiments to find the direction of the magnetic field around a current. (The direction of a magnetic field is given by the direction that a compass needle points when the compass is in the field.) Iron filings may be useful to you. They behave like tiny compass needles in a magnetic field.

- After Oersted's discovery, scientists began to look for the opposite effect: Could electricity be generated from a magnet? Michael Faraday was convinced that this should be so, but he was unable to generate any electricity by placing even very strong magnets near coils of wire. Then, on Christmas Day in 1831, he found the effect he had been looking for.

  You can share his discovery by using long wires to connect a galvanometer to the ends of a coil of wire as shown in Figure 9. Watch the galvanometer needle as you move a strong bar magnet in and out of the coil. You have seen the effect that Faraday discovered. Only when the magnetic field is changing do we find evidence of an electric current.

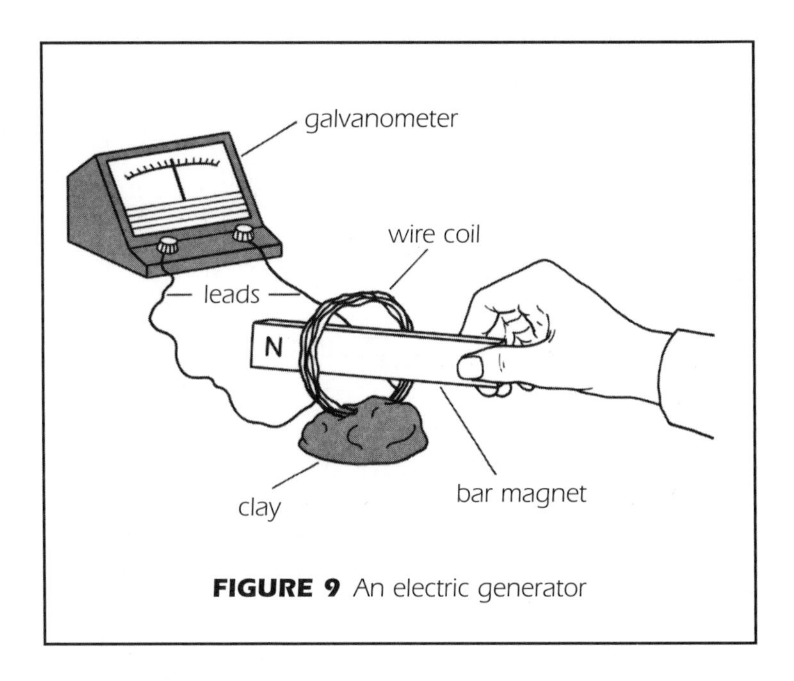

FIGURE 9 *An electric generator*

- Today, with our understanding of the conservation of energy, Faraday's discovery makes sense. We would never expect to see electrical energy generated without some decrease in another form of energy, but Faraday did not know about the law of conservation of energy.

See if you can build a better electric generator, one that will light a flashlight bulb or turn a small electric motor. What forms of energy can you use to produce electrical energy from your generator?

CHAPTER 5

ENERGY OF MOTION AND POSITION

To scientists, work has a somewhat different meaning than it does to most people. Scientists define work as the product of a force and the distance through which that force acts in a direction parallel to the force.

WORK, KINETIC ENERGY, AND POTENTIAL ENERGY

If you pull a cart 10 m along a level floor with a net force of 100 N in a direction parallel to the floor, the work you do on the cart is 1,000 N-m, or 1,000 J. As you do this work, you transfer energy to the cart. The motion that the cart acquires is called kinetic energy. You can see from Figure 10 that the kinetic energy the cart acquires ($\frac{1}{2}mv^2$) is equal to the work ($F\Delta d$) done on the cart.

You actually did more than 1,000 J of work in pulling

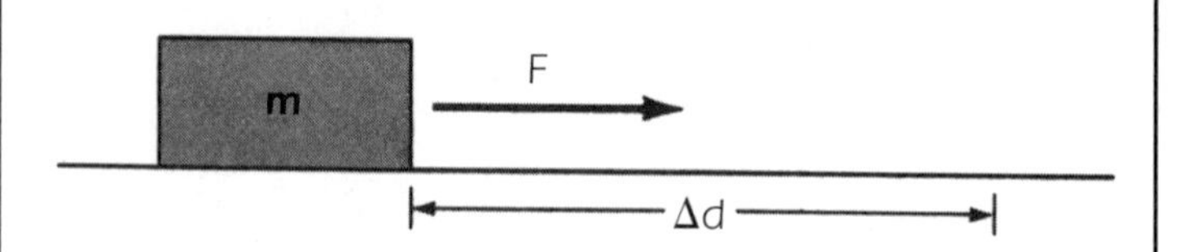

Work done = $F \Delta d$

Since $F = ma$, work = $ma\Delta d$

But $a = \frac{v_f - v_i}{t}$, where v_f is the final velocity, v_i is the initial velocity, and t is the time during which v_i changed to v_f.

And $\Delta d = \left(\frac{v_f + v_i}{2}\right) t$ since distance traveled = average velocity x time.

We conclude that:

$$\text{Work} = F\Delta d = ma\Delta d = m\left(\frac{v_f - v_i}{t}\right) \times \left(\frac{v_f + v_i}{2}\right) t$$

$$= \frac{m}{2}(v_f^2 - v_i^2) = \frac{1}{2}mv_f^2 - \frac{1}{2}mv_i^2$$

The kinetic energy of a moving mass is defined as $\frac{1}{2}mv^2$. As you can see, the work done on the mass equals the change in its kinetic energy.

FIGURE 10 Work and kinetic energy

the cart because there was friction between the wheels and the floor. The net force on the cart was 100 N, but you may have been applying a force of 105 N in order to overcome the 5 N of force between the wheels and floor. You did 1,050 J of work (105 N $\times$ 10 m), but only 1,000 J was transferred into the kinetic energy of the cart; the remaining 50 J has been transferred into thermal energy in the wheels and floor. You do work whenever you lift a mass, because every object near the earth's surface is pulled toward the earth's center with a force of 9.8 N for every kilogram of mass in the object. This 9.8-N force per kilogram that the earth exerts on any mass, m, is called gravity, g. Hence, the weight of any mass, m, is $m \times g$, or mg. Since any mass is pulled toward the earth with a force of 9.8 N, you have to exert an equal force upward to prevent its fall.

If you lift a 1-kg mass to a height of 2 m, you do 19.6 J of work (9.8 N $\times$ 2 m). If that mass is then placed on a shelf, it can, if it falls, do 19.6 J of work by virtue of its position. By lifting the mass, you have stored energy in it. We call the energy stored in an object because of its position relative to the earth's surface *gravitational potential energy*.

In general, the gravitational potential energy (*GPE*) of an object is given by the product of its weight (mg) and its height (h)

$$\text{GPE} = mgh$$

PROJECT 19: THE SPEED OF A FALLING BODY

What You Need	
2 balls, with different masses	Electronic timer connected to a computer
Paper tape timer (with tape and carbon disks) or	Stopwatch

Simultaneously drop two balls that have different masses from the same height. You will see that both balls reach the floor at the same time. They accelerate at the same rate under the forces of gravity. In fact, you can prove to yourself that all masses undergo the same acceleration near the surface of the earth if air resistance is negligible.

What happens to the gravitational potential energy of a mass as it falls? What happens to its kinetic energy? Can you predict the speed of a mass after it has fallen various distances? To measure the speed, you can use a vibrating timer through which the falling mass pulls a narrow strip of white paper. A rotating piece of carbon paper beneath the timer's vibrator will leave a dot on the paper at equal intervals of time. (In many timers the time interval between vibrations is 1/60 s.) Or, with the help of an adult, you could use special spark paper tape that the falling mass pulls through a spark timer that generates an electric spark every 1/60 s. Other ways to measure speed would be to use electronic timers that you could place at different positions along the path of the falling mass, or a computer interfacing device. Check with Vernier Software at http://www.vernier.com for details.

Figure 11 shows how to measure speed at a particular place on a tape pulled through a vibrating or spark timer.

What happens to the speed of the mass as it falls?

Niagara Falls in Upstate New York. Falling water such as this is a tremendous source of energy.

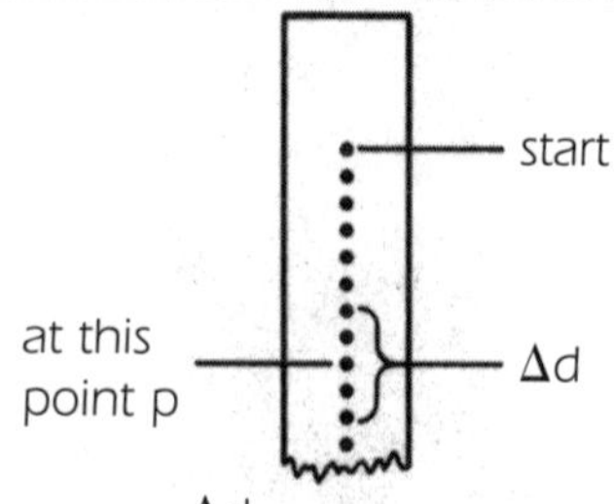

This $\frac{\Delta d}{\Delta t}$ gives the approximate instantaneous speed at point p, which might be 10cm from the start. If $\Delta d = 4.0$ cm and we see that $\Delta t = 4 \times \frac{1}{60}\text{ s} = \frac{1}{15}\text{ s}$, then $v = \frac{\Delta d}{\Delta t} = \frac{4.0\text{ cm}}{\frac{1}{15}\text{ s}} = 60$ cm/s

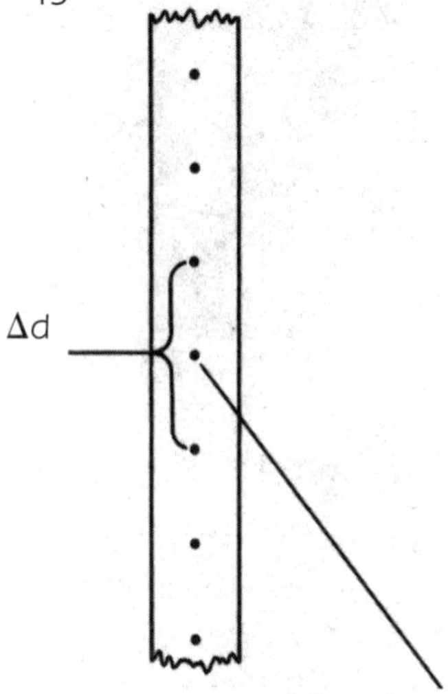

Δd

At this point p′, much farther along the tape, say, 80 cm from start, v (at p′) = $\frac{\Delta d}{\Delta t}$. If $\Delta d = 6.0$ cm and $\Delta t = 2 \times \frac{1}{60}$ s, as you can see, then at p′, $v = \frac{\Delta d}{\Delta t} = \frac{6.0\text{ cm}}{\frac{1}{30}\text{ s}} = 180$ cm/s.

FIGURE 11 Measuring speed using "ticker tape"

Were you able to predict its speed after it had fallen various distances?

Plot a graph of the speed of the mass versus the height through which it had fallen. Is its speed proportional to the distance it fell? Use *Graphical Analysis* (from Vernier) to make the job easier.

Try a graph of speed squared versus the height through which the mass fell. What did you find? Use *Graphical Analysis*' automatic curve fit command and choose "linear" and then "exponential" fits. Which works better? Try some of the other choices of mathematical families. Do any give you a more accurate fit to your data? Can you determine why certain curves work better? How does the kinetic energy acquired by the falling mass compare with the gravitational potential energy lost during the fall?

PROJECT 20: THE KINETIC ENERGY OF A BASEBALL

What You Need	
Baseball	Stopwatch
Bat	

When you throw a baseball, you exert an average force on the ball through some distance. You therefore do work on the ball. The work that you do becomes the ball's kinetic energy. To determine the kinetic energy of the ball, you will need to know both its horizontal velocity (v_x) and its vertical velocity (v_y) at the time you release it. Once these are known, the kinetic energy, K.E., can be determined using the Pythagorean Theorem,

$$\text{K.E.} = \tfrac{1}{2}\, mv^2 = \tfrac{1}{2}\, m\,(v_x^2 + v_y^2)$$

The ball released by a pitcher's arm (in this case that of former New York Met Mike Hampton) has considerable kinetic energy.

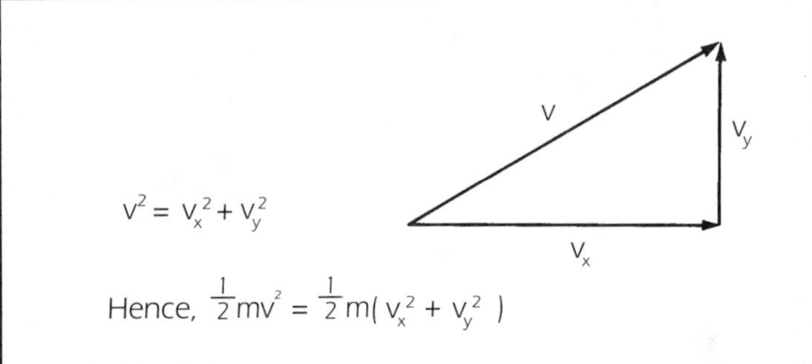

FIGURE 12 Determining the velocity of a thrown baseball

Once the ball is released, its horizontal velocity remains constant because the only force on the ball is gravity, which has no horizontal component.

Have someone use a stopwatch to determine the total time (t) the ball is in the air. By measuring the distance (x) it travels horizontally, you can calculate its horizontal velocity because $v_x = x/t$.

The ball's vertical velocity can be determined because all objects near the earth's surface accelerate downward at 9.8 m/s^2; hence, the time it takes the ball to reach its peak height will be the time it takes for the ball's vertical velocity to change from v_y to zero. Since acceleration is the change in velocity (Δv) divided by the time,

$$\Delta v/t = g \text{ or } (v_y - 0)/t = g \text{ and } v_y = gt$$

The time it takes for the ball to reach the peak of its path will be half its total time in the air; therefore, you can determine the ball's vertical velocity by multiplying half the ball's time in the air by g. What was the kinetic energy of the ball you threw? How much work did you do on the ball? If you can find a way to measure the distance your arm moves as you throw a ball, you can determine the average force that you exert on the ball. How does it compare with the average force that others exert?

Knowing the mass and velocity of the ball, you can easily find the ball's momentum at its point of release. Since the momentum the ball acquires is the product of the average force and the time that force was exerted, you can also determine the time that it took you to throw the ball.

$$m\Delta v = f\Delta t \ (f\Delta t \text{ is defined as "impulse"})$$

What simplifying assumptions have been made in this experiment?

Knowing the change in velocity, mass of the occupant, and the stopping time during a collision, that same equation above can be used to determine forces on car occupants during car crashes.

Early people of the Americas used a spear-throwing device called an atlatl. It was a stick with a curved end that fit into a shallow cup on the end of a spear. This basically made the thrower's arm longer and so a much greater force was generated during spear throws. Did such a device increase the energy of the spear? Can you make a similar "arm-lengthener" for throwing a baseball and check for velocities and energies?

Doing More

- Design and conduct an experiment to determine the kinetic energy of a batted ball or one hit with a fungo bat.

PROJECT 21: ELASTIC POTENTIAL ENERGY

What You Need	
Spring (see specifications below) Clothespins Metric ruler or meter stick	Kilogram mass, other masses (700 g, 500 g, 200 g, etc.) Stopwatch

Choose a spring that increases in length by about 0.3 to 0.5 m when a 1-kg mass is hung on it. To such a suspended spring attach a kilogram mass and support it so the spring is not stretched. Then release the mass so it falls, stretching the spring as it goes. Mark the points of zero and maximum stretch as shown in Figure 13. Notice that the contracting spring returns the mass to very nearly the point from which it was released. This shows that the gravitational potential energy that the mass lost is stored in the stretched spring in a form of energy called elastic potential energy.

To find a relationship between the elastic potential energy and the stretch of the spring, measure the maximum stretch of the spring when various masses (1.0, 0.7, 0.5, and 0.2 kg) are released while attached to the unstretched spring. From the gravitational potential energy lost by the mass, you can determine the elastic potential energy acquired since they are virtually the same in each case.

Plot a graph of the elastic potential energy stored in the spring for each mass that fell versus the stretch of the spring. Is the elastic potential energy in the spring proportional to the spring's stretch? If not, plot elastic poten-

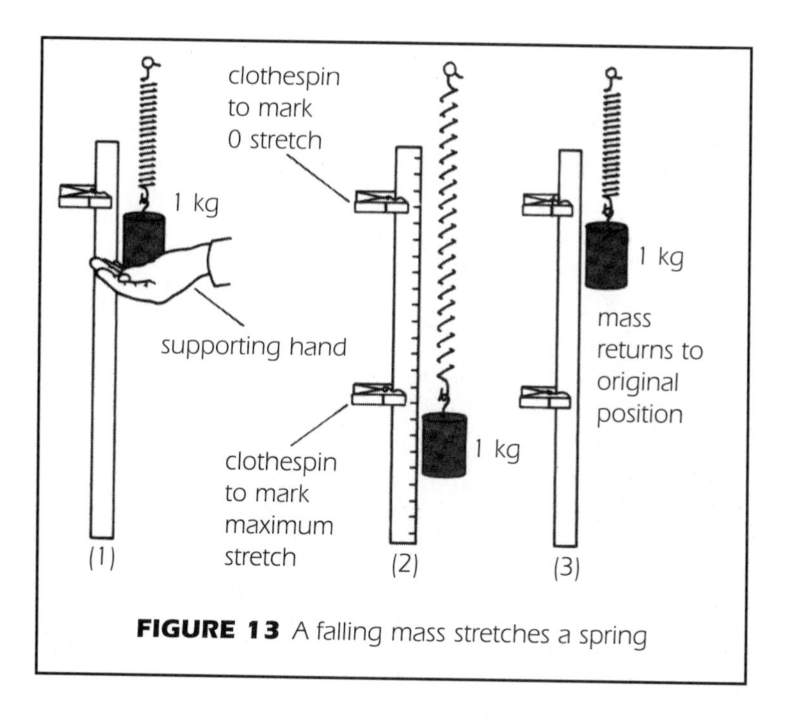

FIGURE 13 A falling mass stretches a spring

tial energy versus the square of the spring's stretch. What does the graph tell you? Write the equation for the graph. Use *Graphical Analysis* to help in graphing and finding the best equation type for your data.

Doing More

- You may have noticed that when you first dropped the masses, they bounced up and down for a while. Use the stopwatch to time the period of each bounce of the spring as it is coming to rest with each of your masses. Graph the results with *Graphical Analysis* and get an equation for your work. Repeat the experiment many times to get a reliable curve. Now, using your equation, predict the period for a new mass you haven't tried before. How accurate was your prediction? Investigate the mechanism of a wind-up clock (try books such as *How Does it Work* or find an old junk alarm clock and take it apart). How does the mechanism relate to your experiment? Could your experiment be used to tell the mass of an object in a place of zero gravity? How would you have to modify it?

- To check up on the relationship you have just found, you can measure the stretch (increased length of the spring), in meters, versus the force acting on the spring. Mark the bottom of the spring when it is not stretched at all. Then attach a 0.20-kg mass to the spring and carefully lower the mass until the spring stops stretching. Measure the increased length of the spring when this force of 1.96 N pulls on the string. Repeat using forces of 4.9, 9.8, and 14.7 N by hanging masses of 0.50, 1.00, and 1.50 kg from the spring. Record the increased length of the spring for each force.

  Now plot a graph of the force applied to the spring versus the stretch of the spring. Is the stretch proportional to the force applied? Use *Graphical Analysis* to see how close your data fits the theoretical equation of $f = kd$, where k is some spring constant and d is the distance stretched or compressed.

  Notice that the area under the graph of force as a function of stretch measures the work done on the spring or the potential energy stored in it. Write the equation for the graph you have just drawn. How do the results of this experiment compare with those in Project 21?

- The gravitational potential energy lost when a mass falls while attached to a spring has changed to elastic potential energy when the mass stops falling. The elastic potential energy is then changed back to gravitational potential energy as the mass ascends to its original position. But during the transition from one form of potential energy to another, there exists not only gravitational and elastic potential energies, but kinetic energy as well.

Recognizing that energy is conserved, try to predict how much kinetic energy should exist at various points in the oscillation of a mass on a spring. Then determine by an experiment of your own design whether your predictions are correct.

- When molecules of gas collide, the collisions are generally elastic; that is, the total kinetic energy of the molecules before and after the collisions is the same. However, collisions of most common objects are inelastic because some kinetic energy is lost during collision. In an elastic collision, some kinetic energy is converted to potential (often elastic potential) energy as the colliding bodies approach, but the potential energy is returned to kinetic as the bodies separate. Inelastic collisions, such as the collisions of cars, are evident by the permanent change in structure. The metals dented or bent in the collision are not sufficiently elastic to return to their original shapes; consequently, the work done in compressing materials as the objects approach one another is never returned. The result is crumpled, permanently damaged matter.

  From what you have learned about elastic potential energy and collisions, and the equation at the end of Project 20, see if you can design a better bumper for automobiles that will resist permanent deformation and make collisions more elastic. Could you store the energy in *magnetic* fields? As you work on your design, think about the occupants of such cars. Will your design provide greater safety to the passengers?

PROJECT 22: POTENTIAL ENERGY IN MOLECULES

What You Need	
Steam generator and trap (see Figure 14)	Glass tubing (about 6 mm in diameter)
Bunsen burner setup	Rubber tubing (to fit glass tubing tightly)
Foam cup	Thermometer or temperature probe
Cold water	
Balance (scale)	
Graduated cylinder	

When you boil a liquid, the temperature of the liquid does not change despite the fact that thermal energy is being added to the liquid. However, the liquid does disappear as it changes to a gas. Careful measurements show that the temperature of the gas is the same as that of the boiling liquid. Could it be that the separation of the molecules as they change from liquid to gas results in potential energy among the molecules, just as the separation of a mass from the earth leads to an increase in gravitational potential energy?

To test this idea, see what happens when a gas condenses to form a liquid. If potential energy increases when molecules separate, then we might expect a decrease in potential energy when the gas molecules coalesce. This decrease in potential energy should result in the release of an equal amount of another form of energy—probably the kinetic energy of molecules, which we call thermal energy.

Use a steam generator and trap as shown in Figure 14 to produce steam. (BE CAREFUL. STEAM CAN BURN YOU TEN TIMES MORE SEVERELY THAN BURNS FROM BOILING WATER AT THE SAME TEMPERATURE!) You can then condense some of the

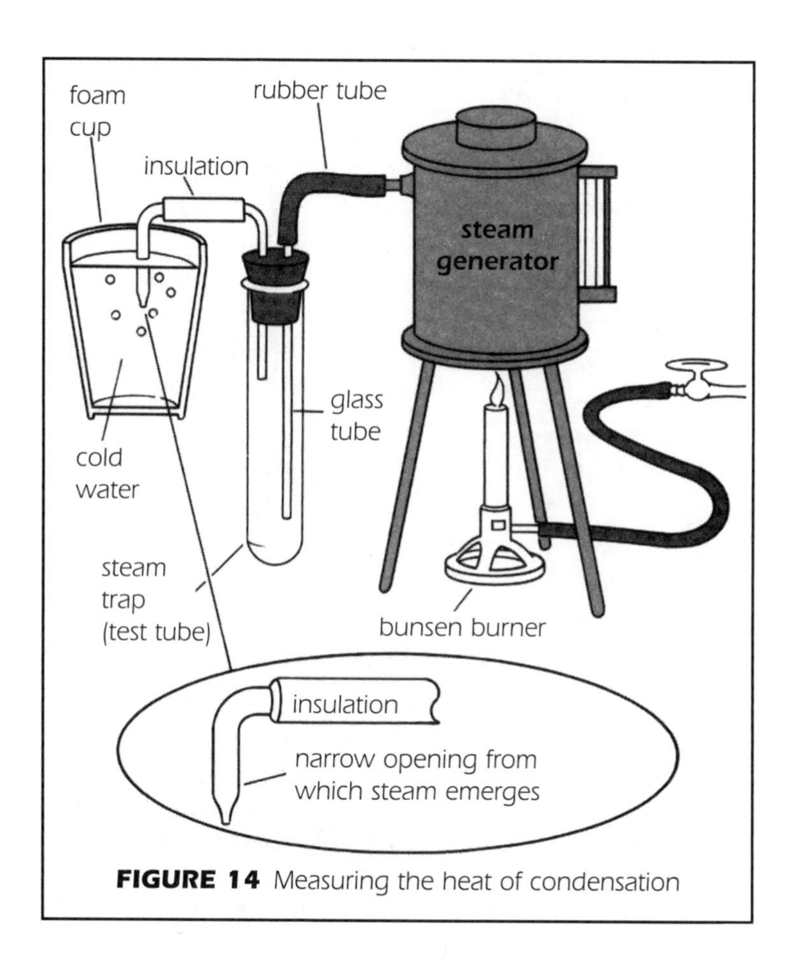

FIGURE 14 Measuring the heat of condensation

steam in cold water and see how much heat is transferred per gram of steam condensed.

Fill a weighed, foam cup about three-fourths full with cold water about 10°C below room temperature and reweigh. Determine the temperature of the water. Then place the tapered glass tube from which steam is emerging into the cold water. Allow steam to condense in the water until its temperature rises to about 10°C above room temperature. Remove the water from the steam source and record its final temperature. Then reweigh to find out how much steam is condensed. Why should you let the temperature of the cold water rise until it is about as many degrees above room temperature as it was below room temperature when the experiment started?

The temperature increase of the cold water will tell you how much heat was transferred; however, not all that heat came from the condensing steam. Remember, after the steam has condensed, it is at the temperature of the boiling water (about 100°C). The condensed steam loses heat in cooling to the final temperature of the water. By subtracting this heat from the total heat transferred to the cold water, you will be left with the heat that came from the condensation of the steam. How much heat was released per gram of steam that condensed?

Is there evidence to support the idea that potential energy is stored in molecules when they separate?

Doing More

- Using your investigation as a starting point, create a detailed report on the physics of the espresso machines used by coffee shops. What causes that hissing sound? How much energy is used to make the typical cup of espresso? How much of a temperature change is there in the average cup? How much liquid undergoes a temperate change? Can you use these data to determine exactly how much steam gets delivered? If you really want to win first-place at the science fair, set up an exhibit and serve coffee to the judges.

PROJECT 23: HEAT OF VAPORIZATION

What You Need	
Immersion heater (type used to heat a single cup of water)	Cold water
Foam cup (12-oz. size)	Ice (for "Doing More" only)
Graduated cylinder	Stopwatch
Thermometer or temperature probe	

It is more difficult to measure the heat of vaporization (the heat required to change a gram of liquid to vapor at the boiling point) than the heat of condensation because it is not easy to compensate for the large heat losses to the surroundings that occur at 100°C.

To obtain an approximate value, first calibrate an immersion heater (the type sold to warm just one cup at a time). Place about 150 g (150 ml) of cold water in a 12-oz. foam cup. This larger size cup will reduce water losses due to spattering when the water boils, and the insulation will reduce heat losses. Measure the temperature change of the water when the immersion heater is used to heat the water for a period of 1 min. You may find that a temperature probe connected to a computer will give you a faster reading of the temperature changes than a glass thermometer. Remember to stir the water so the probe or thermometer measures the overall temperature, not just a hot spot. NEVER CONNECT AN IMMERSION HEATER TO AN ELECTRICAL OUTLET UNLESS THE HEATER IS IN WATER.

How much heat does the heater deliver in 1 min? If you use a temperature probe and have it record temperatures every second, you can get a heating rate of degrees per second of operation.

Use the same heater to bring 150 g of water to boiling. Let the water boil for several minutes so that some of the water is converted to gas (steam). At a convenient time, disconnect the heater and carefully pour the water that remains into a graduated cylinder. Let the water cool and measure it again. Did the contraction of the water when cooling cause a significant volume change? How much liquid water was changed to gas? How much heat was added to the water? How much heat was required to boil away 1 g of water? (Remember to subtract the heat that was required to bring the water from its initial temperature to the boiling point.) How does your value for the heat required to boil away a gram of water compare with the heat released when a gram of steam condenses? How would you expect these values to compare? Design an experiment to reduce or compensate for heat losses and thus obtain a better value for the heat of vaporization of water.

Doing More

- The heat required to melt 1 g of solid at its melting temperature is called the heat of melting. Design and carry out an experiment to determine the heat of melting for ice.

The heat of fusion is the heat that must be removed from 1 g of a substance to change it from liquid to solid at its freezing temperature. How would you expect the heat of fusion and the heat of melting of the same substance to compare? Design an experiment to measure the heat of fusion for water. You may find it more convenient to use a temperature probe than a thermometer since it can be frozen in place and still be read.

PROJECT 24: THE HEAT OF FUSION FOR HYPO

What You Need	
Test tube	100 g sodium thiosulfate pentahydrate [$Na_2S_2O_3 \bullet 5H_2O$]
Beaker	100 g sodium sulfate decahydrate [Glauber's salt, $Na_2SO_4 \bullet 10H_2O$]
Foam cup	Stopwatch
Lab burner	
Thermometer or temperature probe	

See if you can determine the heat of fusion for sodium thiosulfate [$Na_2S_2O_3 \bullet 5H_2O$], known to photographers as "hypo" or "fixer." Place a test tube half-full of the crystals in a beaker of hot water to make them melt. This substance freezes at 48°C, but it is easily supercooled to 35°C or lower. So it is best to cool the solution to about 35°C and then add a tiny crystal of the substance, called a seed crystal, to initiate the freezing process.

After it freezes, let it cool back to 35°C before making any final measurements. Any heat required to raise the hypo from its supercooled temperature to its freezing temperature will be returned when it cools back after freezing. Cool water in a foam cup can be placed around the test tube of hypo to measure the heat of fusion once the hypo starts to freeze.

Glauber's salt [$Na_2SO_4 \bullet 10H_2O$] is a substance that has been used to store solar energy, so you might like to measure its heat of fusion for future use if you do solar energy projects. In storing solar energy as hot liquids or solids, it is best to have the storage medium take up as little space as possible for the practicality of use in small homes or apartments. Using the density of both sodium thiosulfate and Glauber's salt, and using the data from your measurements of the calories of energy for melting, which one would be the most efficient in kcals per volume?

Doing More

- Contact a local fuel company or electric utility and investigate how many therms, Btus, or kcals of energy are needed to keep the average house in your area warm from sunset to sunrise. Now use your data from the previous experiment to calculate both the size and the weight of a tank needed to hold enough sodium thiosulfate or Glauber's salt to contain that much energy overnight before it is heated by the next day's solar energy.

CHAPTER 6

SOLAR ENERGY

The source of the chemical energy stored in fossil fuels, the source of the energy that drives windmills, and the source of the energy that lifts water into the atmosphere from which it can fall and flow over dams, providing the kinetic energy to turn the turbines that generate electricity are all the same. The source of all these energies is solar energy, the energy that comes to us from the sun. It is the earth's only outside source of energy, and it is free.

Solar energy causes none of the pollution associated with the burning of coal and oil, and we do not have to tear the earth open to find it. Sunlight falls on every nation of the world and cannot be controlled or hoarded by select nations or terrorists.

Unfortunately, solar energy is diffuse—there is relatively little energy per unit area. The amount in even the sunniest places, such as Arizona, is only about 1 horsepower per square yard. Consequently, to use solar energy efficiently, we must build devices that concentrate the light into smaller areas or make collecting devices that have large surface areas.

PROJECT 25: HEATING AIR WITH SOLAR ENERGY

What You Need	
2 or more aluminum pie pans Black paint 2 thermometers or temperature probes	Tape, plastic bags, cardboard box, soda straw, long pin, protractor
For the "Doing More" section, you will need	
Graduated cylinder Water Black ink	Colored construction paper (various colors) Stopwatch

To see how solar energy can be used to heat air, first find two identical aluminum pie plates. Paint the inside of one with flat black paint. When the paint has dried, fix a thermometer to the edge of each pan with masking tape. Be sure the tape covers the thermometers' bulbs so that the sunlight does not fall directly on the glass bulbs. Place each pan and thermometer inside identical clear plastic bags. Seal the bags with twist-ties. Place both pans in a box and position the box so that sunlight is perpendicular to the surface of the pans.

Check the air temperatures in the bags periodically

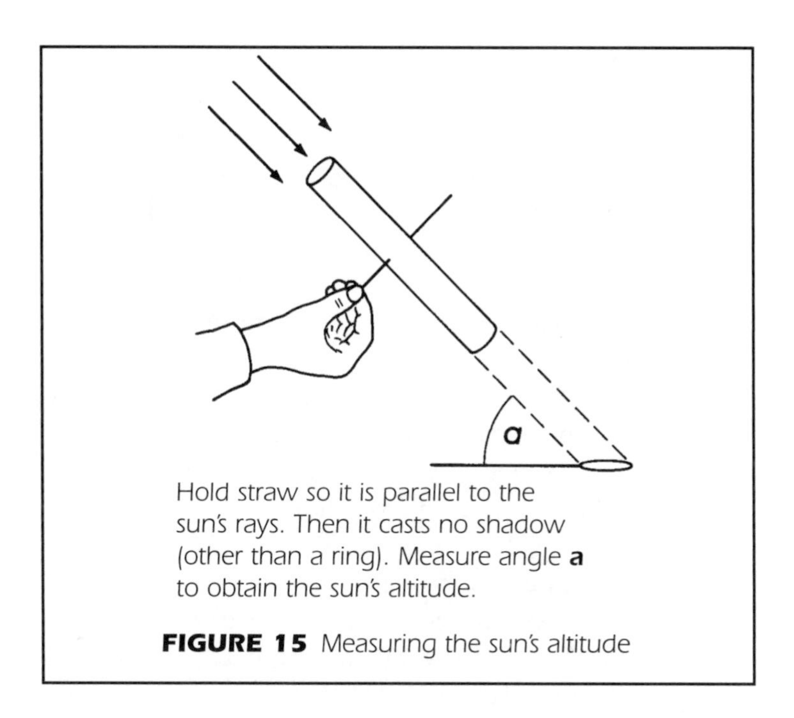

Hold straw so it is parallel to the sun's rays. Then it casts no shadow (other than a ring). Measure angle **a** to obtain the sun's altitude.

FIGURE 15 Measuring the sun's altitude

until the temperatures stop rising. Which pan converts more solar energy to heat?

Repeat the experiment using a large and a small pan both painted black. Which pan produces more heat?

This time use several black pans. Vary the angle that each pan makes with the sun's rays. To find a surface that is perpendicular to the sun, mount a soda straw on a pen and hold it above a flat surface as shown in Figure 15. When the straw casts no shadow of its sides, you know it is parallel to the sunlight. Use a protractor to measure the angle that the sun's rays make with the ground. How can you determine the angle between the sun's rays and the surface of each pie pan tilted at different angles?

Which angle seems to produce the maximum amount of heat from solar energy?

Doing More

- Water, too, can be warmed by the sun's energy. To see how this is done, place 100 ml of water in each of two aluminum pans. One of the pans should be painted black as in Project 25; the other should not be painted. Place both pans on an insulating sheet of cardboard in a warm, sunny location. How can you determine which pan will convert more solar energy to heat?

  Repeat the experiment, but this time use two unpainted pie pans. Add enough black ink to one pan to make the water very dark. Which pan do you think will produce more heat from sunlight?

- Does the color of a house or roof have any effect on the solar energy reflected or absorbed? To find out, cover the bulbs of several thermometers with small sheets (5 cm × 10 cm) of colored construction paper. You might choose red, green, blue, white, black, and any other colors that you may want. Fold the paper around each thermometer bulb in the same way. Use a paper clip to hold the paper in place.

  Note the temperature within each colored paper before placing the thermometers side by side on an insulating sheet of cardboard in a bright, sunny place. Record the temperature within each colored sheet at 1-min intervals for a period of about 15 min. Which color seems to be best suited for converting light to heat? Which one seems to be the best reflector of light?

  In the winter, you might do this experiment by placing pieces of metal, identical in every way except for their color, on the surface of some snow. How can you determine which piece of metal is converting the most sunlight into heat?

PROJECT 26: THE GREENHOUSE EFFECT

What You Need	
Small thermometer Glass jar Piece of cardboard (see Figure 16)	Stopwatch
For the "Doing More" section, you will need	
Insulating material of various types Protractor Wood, glass, and plastic to construct a model greenhouse Large cardboard box Newspaper 2 funnels, rubber tubing, collecting cup	Aluminum foil (heavy duty) Flat black paint Soda straw, clay Graduated cylinder Small pump (optional) Plastic wrap and tape

The place in which you have most likely seen the greenhouse effect, strangely enough, is not in a greenhouse but in a car. If you have ever stepped into a car that has been tightly closed and sitting in the sun for some time, you have undoubtedly noticed how much warmer it is inside the car than outside. This warming of a glass-enclosed structure sitting in sunlight is called the greenhouse effect. The sun's rays pass through the glass, but cannot pass back out. In a similar manner, the earth's atmosphere reflects solar radiation back to the ground, creating a greenhouse effect on a global scale. Water vapor is the

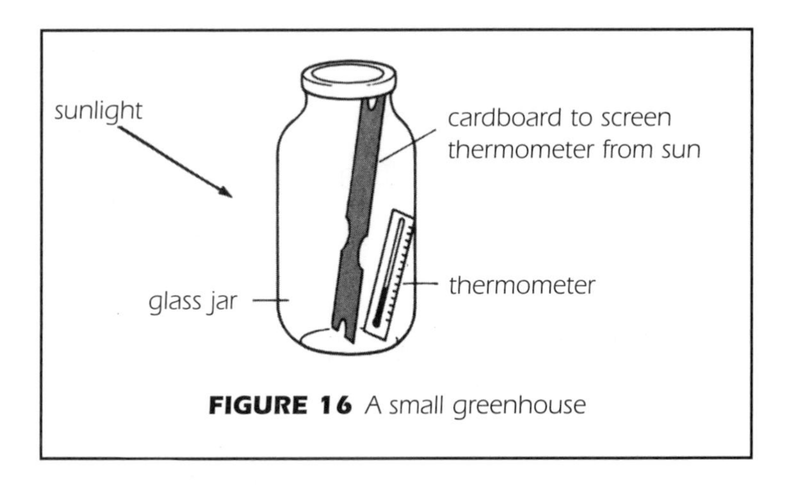

FIGURE 16 A small greenhouse

primary "greenhouse gas" for earth, although CO_2 and methane also play a part.

To get a closer look at this effect, place a small thermometer inside a glass jar. Insert a piece of cardboard as shown in Figure 16 to be sure that the thermometer bulb is shaded. Place the jar in bright sunlight with its sun-facing side perpendicular to the sunlight. Watch the temperature in the jar for a few minutes. What happens?

Can you get the temperature to rise higher by covering the top, bottom, and back side of the jar with insulation?

Does the angle that the sun-facing side of the jar makes with the sunlight have any effect on the rate at which heat is produced in the jar?

From what you have learned, see if you can design and build a model greenhouse that will collect solar energy. If you are successful, perhaps you can design and build a greenhouse on the south side of your house that will serve as an additional heat source and reduce your family's heating costs. At the same time, you will have a warm room for winter living as well as a place to grow plants year-round.

Doing More

- You have probably seen solar collectors on the south-facing roofs of a number of homes and commercial buildings. These collectors are usually used to heat water. If, as is usually the case, pumps are used to move water between the collectors and the tank where the hot water is stored, the collectors are referred to as active solar collectors. Generally, there is a backup system to ensure the presence of hot water if the weather is cloudy for several days. Occasionally, active solar collectors are used to heat air in buildings as well, but more commonly passive methods—methods that do not involve the use of electrical devices such as pumps—are used to heat the space within buildings.

  You can build a small model solar collector in the following way. First, cut the flaps from a cardboard box (about 20 cm × 30 cm × 10 cm deep). Fill the lower half of the box with crumpled newspaper for insulation. Punch a hole in one corner of the box about 3.8 cm from the top and one side as shown in Figure 17. Push the spout of a funnel through the hole. The funnel should rest in one corner of the box and serve to collect water that will flow down the surface of the collector.

  Next, cut a sheet of cardboard to the same width as the box. Its length should equal the distance from the top of the funnel to the other end of the box. The fit should be a snug one to help hold the funnel in place. Cut a piece of heavy-duty aluminum foil about 5 cm wider than the box with a length equal to that of the box. Paint the aluminum with flat black paint.

  Once the paint is dry, lay the foil on the cardboard. Fold the "extra" foil against the sides of the box and tape it firmly in place. Turn up the bottom

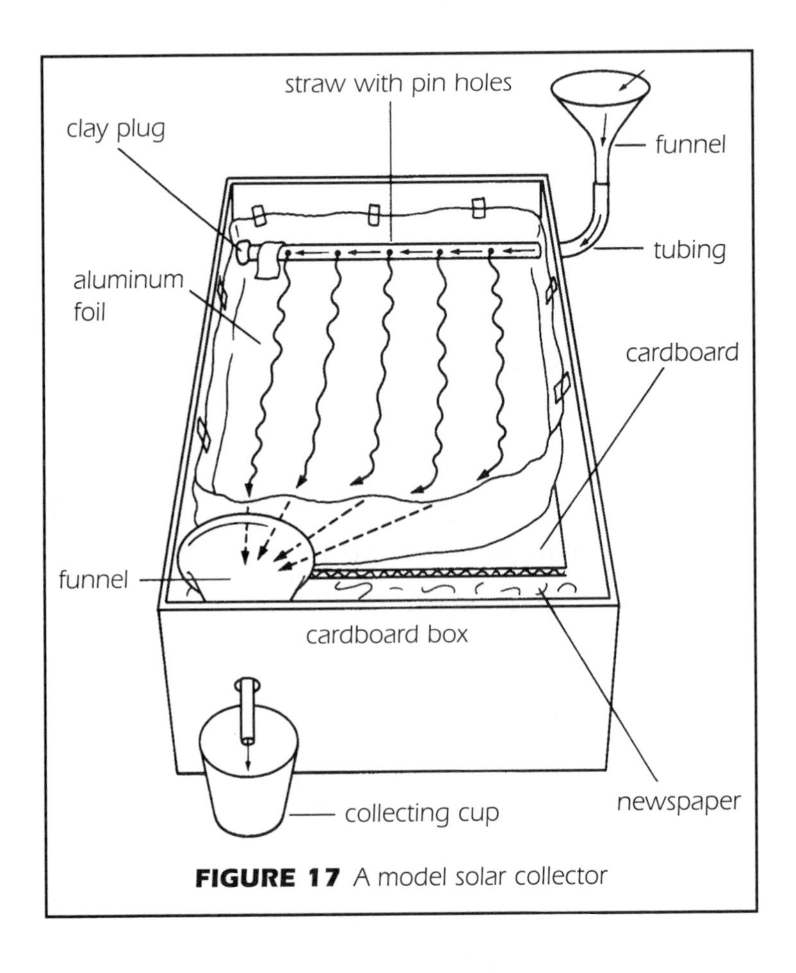

FIGURE 17 A model solar collector

edge of the foil to form a trough that will carry water into the funnel. Punch a hole through the foil above the funnel so that water can run out of the box into a collecting cup. Use a thick pin to punch six to seven holes in a straight line along one side of a giant (0.8 cm in diameter) plastic soda straw or tube. Punch a hole in the upper corner of the box and push the straw through the hole. The straw should rest on the aluminum foil with its holes near the foil's surface. A clay

Solar-powered pay phone in Red Center, Australia.

plug over the end of the straw within the box will force water to flow out of the holes. Use tape to fasten the straw firmly against the aluminum foil. The other end of the straw protrudes through the box and is connected to another funnel by a short piece of rubber tubing.

Place the box in a sunny place with the aluminum foil facing the sun. After measuring the temperature of 50 ml of cold water, pour the water into the upper funnel. The water will flow into the straw, out the small holes, and down the warm aluminum foil to the trough, where it will collect and run into the funnel and out of the collector into a cup.

In a real collector a pump would return the water to the top of the panel after it had passed over the warm metal surface and into a storage tank. In this model collector you will have to remove the cup and pour the water back into the upper funnel so that the water may again flow over the metal foil. Be sure to put another cup under the lower funnel. On the other hand, you might add a pump. Plant stores and crafts shops sell very small pumps for tiny indoor fountains at a low price, complete with plastic tubing.

After the water has flowed over the collector surface a number of times, measure the water temperature again. How much has the temperature risen? How much warmer can you get the water by repeating the process again? Measure the temperature of the water using a temperature probe and graph the collected data vs. time using *Graphical Analysis*. Is the result a straight line or a curve? What explanation can you offer for the graph's shape?

Cover the open side of the box with plastic wrap and tape the clear material to the sides of the box. Repeat the experiment. Why does the water heat up faster this time? Now see if you can build a better model solar collector. As you become more and more familiar with these devices, you may want to build a collector that can actually be used to heat water in your house. Will you use water or air to transfer heat?

PROJECT 27: WHICH WAY IS SOUTH?

What You Need	
Stick String	Watch Straight piece of board
For the "Doing More" section, you will need	
Clear plastic dome or fine-mesh hemispherical kitchen strainer Heavy cardboard, tape	Marking pen Colored round-headed pins, yarn (if you use the strainer)

To build a solar-heated home in the northern hemisphere, the architect and builder must know which direction is south so that large windows will face the winter sun.

The simplest method is to find the North Star and align it with several sticks to get a north-south line, but this may be difficult to do on a moonless night.

You can obtain a north-south line in another way during the midday hours. At about 11 A.M. standard time on a sunny day, push a stick into the ground. Be sure the stick is vertical. With a string and a sharp stick, scratch a circle around the vertical stick using the shadow of the stick as a radius. Mark the spot at which the shadow touches the circle with a stone. The shadow will continue to move and grow shorter as midday approaches. After midday (this does not necessarily occur at noon, depending on where you are in your time zone), the shadow will grow longer and move eastward. Finally, it will touch the circle again. Mark this point with another stone.

Next, draw a chord of the circle by drawing a straight line between the two stones. A line from the stick to the midpoint of the chord will be a north-south line. Explain why this method works, and also explain why a magnetic compass is not a satisfactory way of finding true north in most places.

Some architects maintain that the solar-collecting windows of a solar house should actually face a little west of south. Do you agree with this?

Doing More

- A solar house must have south-facing windows to collect energy from the winter sun. Unfortunately, these windows will allow light from the summer sun to pass through them as well, and the sun stays above the horizon much longer each day in summer than in winter.

  By knowing the position of the sun at various times of the year, an architect can design the roof overhang so as to enhance winter heating by the sun while shading the window when the summer sun is higher in the sky.

  When you stand in an open space and look upward, you see a hemispherically shaped sky surrounding you. At night, stars glisten like tiny lights set in this huge hemisphere. You are at the very center of what astronomers call the celestial hemisphere. To map the sun's path across the sky, you'll need a clear plastic dome or a fine mesh kitchen strainer to represent the hemispherical sky (celestial hemisphere). Put the dome or strainer on a sheet of heavy cardboard. Mark the outline of its base with a pencil. Then remove the dome or strainer and make an X at the center of the circle you have drawn. The X represents your position at the center of the celestial hemisphere. Put the strainer back in its original position and tape it to the cardboard.

Place the dome outside on a level surface shortly after sunrise. To map the sun's path, place the tip of a marking pen on the dome so the shadow of the tip falls on the X at the center of the circle as shown in Figure 18. Mark this point on the dome with your pen. The pen mark on the dome represents the position of the sun in the sky because it is directly in line with the real sun and the X that represents you.

Make marks like this at half-hour intervals throughout the day. By sunset you will have a map of the sun's path across the sky. If you mark north and south on this dome, you can use the same dome to map the sun at different times of the year.

If you use a strainer, colored round-headed pins can be used to cast shadows on the X. Leave pins in the strainer; they will provide a colorful map of the sun's path. To make a permanent record of the sun's

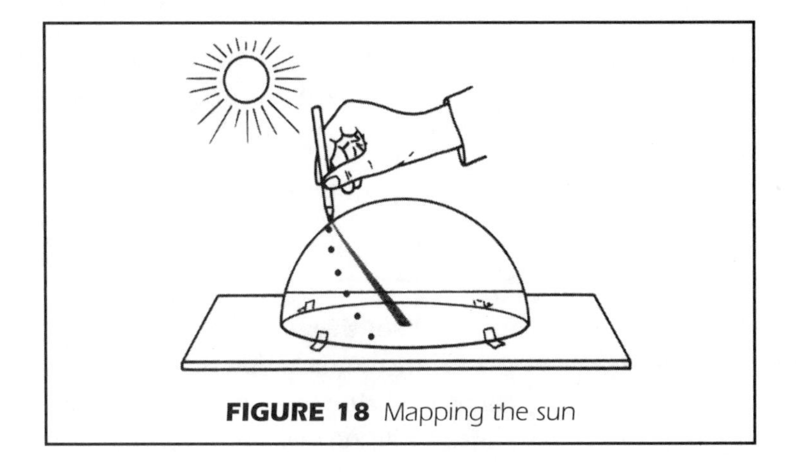

FIGURE 18 Mapping the sun

path, run a piece of colored yarn through the pin's positions as you remove them.

Repeat this experiment at different times of the year. Try especially hard to map the sun on or about the 21st of September, December, March, and June. These dates give particularly interesting maps because they mark the beginning of the seasons. The maps for June and December will give you the sun's maximum and minimum midday altitudes, respectively.

When does the sun rise in the direction due east and set in the direction due west? You can surprise your friends by announcing, "Today the sun rose in the east!" They'll think it does every day, and then you can tell them the results of your research.

How can you use these maps of the sun's path across the sky to determine the altitude of the sun in summer and winter? How will this help you design the proper overhang for a roof on a solar-heated home?

PROJECT 28: STORING SOLAR ENERGY

What You Need	
Small, identical tin cans	Thermometers or temperature probes
Gravel, stones, sand, dirt, salt, water	Flat black paint
Kitchen oven	Stopwatch
Potholders	

A house with plenty of south-facing windows may become so warm on a bright January day that a window will have to be opened; yet, after sunset, an alternative heating system will have to be turned on to replace heat lost to the cooler surroundings. To prevent such wide fluctuations in temperature, there are ways to "bottle" excess daytime solar energy and store it for use at night when outside temperatures plummet.

The Pueblo Indians used thick adobe walls to "soak up" solar energy. Today, many solar home builders use Trombe walls—massive, dark concrete walls—in front of south-facing windows to absorb heat. Air between the wall and window is heated, rises, and passes through slots into the room as cooler air enters through ducts at the base of the wall. During the day the wall absorbs heat, which is then available to warm the building at night. To reduce heat losses, the windows are covered with insulating material after sunset.

To find suitable materials to use for storing solar energy, you can place small identical cans filled with various substances such as gravel, stones, sand, dirt, salt, and water in a 50°C (120°F) oven until all the substances are of that temperature. (Be sure to cover the water loosely so it can't evaporate.)

Using potholders so you won't burn your fingers, remove the substances from the oven, and record their temperatures as they cool. Use temperature probes to get an automatically recorded list over several hours, if needed. Which substance cools fastest? Slowest? Which one would you recommend for storing solar energy?

To test your recommendation, paint the cans with flat black paint. After the paint is dry and the substances are all at the same temperature, place the cans in a sunny place where they all can receive the same amount of sunlight throughout the day. Toward the end of the day, measure the temperature in each can. Which substance has gained the most heat? Is a high temperature the same thing as a lot of heat? Look up definitions. If you measure the temperatures with several probes, you can see which substance increases in temperature the fastest. (You'll need a balance to weigh the materials and an extra empty can. A table of specific heats can be found in Appendix 2.)

Now leave the cans in a room until they all reach room temperature. Which can lost its heat fastest? Which one cooled slowest?

Would you still recommend the same substances to store solar energy?

Doing More

- If you measured the heat of melting of Glauber's salt [$Na_2SO_4 \bullet 10H_2O$] in Chapter 5, you know that energy can be stored in the melted salt as chemical potential energy, or latent heat. If solar energy were used to melt the salt, that energy, which would be released when the salt solidifies, could be used later as a heat source when solar energy is no longer available. A number of people, including Maria Telkes, have experimented with Glauber's salt as well as other salts with low melting points such as sodium thiosulfate pentahydrate [$Na_2S_2O_3 \bullet 5H_2O$]. They claim that these salts can be used to heat buildings in the winter and keep them cool in the summer. (See the appendix for Maria Telkes' book. It is out of print now, but a library may have it.)

  Others have experimented with roof ponds—shallow tanks of water that rest on flat roofs. These ponds absorb solar energy during the daytime. At night they are covered and insulated so that the energy stored in the water can radiate heat into the living space below. During the summer, these same ponds can be used to absorb solar energy so that it does not reach the living space. Then at night, the pond, if left exposed to the dark sky, will radiate the energy absorbed during the daytime.

See if you can design a model system that uses a roof pond or one or more salts with low melting points to store solar energy. Check with your local electrical utility to see how many kcal of heat are necessary to keep a house in your area warm overnight during an average winter night. Ask them about "degree days." To appreciate some building considerations, measure your own house's roof area (assuming that it is flat), then calculate how much the water would weigh if you filled the pond with just five inches of water. One cubic centimeter of water weighs 1 gram; 1,000,000 grams = 1 metric tonne; 1 tonne = 1.1 English tons. Can your roof take that much weight? Is your roof area sufficient for keeping your house warm overnight or would you need extra heating? If you are successful and want to be rich and famous, you might like to build a system that could be used to store solar heat in buildings.

PROJECT 29: TRACKING THE SUN

You've found that the maximum rate of solar energy absorption occurs when the collecting surface is perpendicular to the sun's rays. Therefore, it would be advantageous to have solar collectors turn with the sun and even vary their elevation, if possible, as sunflowers do each day. To date, this has been too expensive in terms of the payback; consequently, most solar collectors are built to face south and are tilted at an angle that makes them reasonably close to being perpendicular to the winter sun's midday light.

See if you can design and build a model for a device that will keep a solar collector facing the sun. If you succeed, you might want to try to build a commercially feasible device. You will probably be able to find financial support from a solar energy company for such a project if your model appears to be practical both mechanically and economically.

PROJECT 30: BUILDING A MODEL SOLAR HOME

Solar homes not only must collect and store solar energy, they must also trap that heat within their walls so that it doesn't escape easily. Consequently, these buildings must be well insulated and weather stripped. With the exception of the south side, there should be few windows, and those should be double or triple glazed and covered at night with insulated shutters or shades.

Now that you've learned something about collecting, storing, and trapping solar energy, you might like to try your hand at building a model solar building that uses passive solar energy as a heating source. What modifications should you make to cool your house naturally during the summer? If your model is a successful one, you might like to apply what you have learned to the design and eventual construction of your own passive solar home.

Doing Less

- If you're not ready to design a solar home, you might like to tackle a less ambitious, but useful, project. Design and build a solar heater that would help warm at least one room in your home. Such a heater could be built just outside a south-facing window. Make sure to use sturdy construction so you don't provide an inviting entrance for burglars!

PROJECT 31: STORM WINDOWS AND DOORS

What You Need	
Small, thick cardboard box (large enough to hold a glass of water)	2 thermometers or temperature probes
Water glass	Access to refrigerator
Sharp knife	Plastic wrap, tape
	Stopwatch

Most well-built homes in northern climates, whether solar-heated or not, have storm windows and doors to reduce heat losses through glass and uninsulated wood.

To see the effect of storm windows and doors, find a small, thick cardboard box just big enough to hold a glass of water. With a sharp knife cut windows on four sides of the box as shown in Figure 19. Fill the glass about three-fourths full of room-temperature water and place it in the box. Tape the top of the box and other seams. Punch a

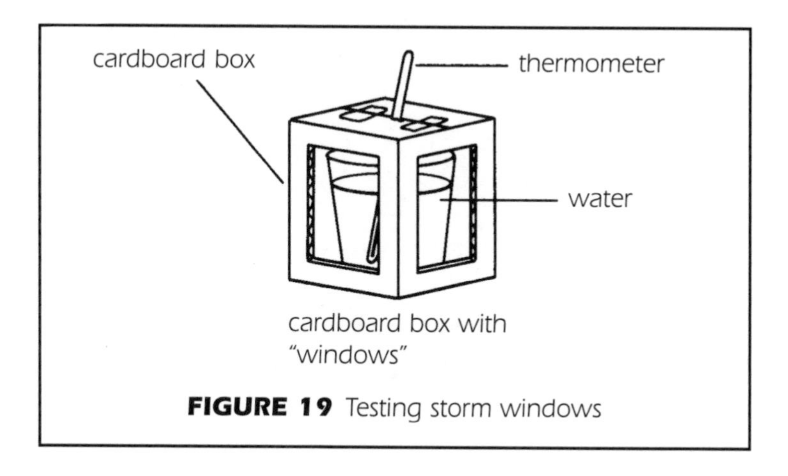

FIGURE 19 Testing storm windows

hole through the top so a thermometer can be pushed through the box and into the water.

Place the box in a refrigerator and record the temperature at 5-min intervals as the box cools through at least 10°C (18°F). Here is where temperature probes work their best. The wires connecting the probes are small enough to fit between the door seal and the refrigerator. Therefore, you can measure and record the temperature without opening the door every five minutes and changing the temperature. Don't forget to put one probe outside your cardboard box to determine the temperature of the refrigerator itself.

Repeat the experiment, but this time tape a layer of plastic wrap over the window on the inside of the box. Pull the wrap so that it is taut and tape it firmly to the inside surface of the box. How does the rate at which heat is lost from the water this time compare with the first run?

Now tape a taut layer of the plastic wrap to the outside of the box. Be sure the two layers do not touch so that there is a layer of air between them. Repeat the experiment. What do you find about the rate at which heat is lost this time? How do you explain your results?

What could you do to further reduce the rate at which heat is lost from the box?

PROJECT 32: SOLAR ENERGY TO MECHANICAL ENERGY

When stretched, polymer plastic strips, which can be made by cutting up polyethylene garbage bags or rubber bands, will contract upon heating. You could, of course, heat them with sunlight. With patience and ingenuity, you might be able to construct a solar-powered motor that could be driven by the conversion of light energy to chemical energy to kinetic energy. If you enjoy tinkering, give this a try. It may not lead to a very efficient engine, but it should be fun.

PROJECT 33: THE SOLAR CONSTANT

The energy that we receive from the sun varies from day to day depending on the season, the amount of cloud cover, air pollution, and other factors. But just above our atmosphere the radiant energy from the sun is nearly constant. Since the energy from the sun is so vast, we usually measure not the total energy but the power per area. We call this value the solar constant. It is 1.95 cal/cm^2 or 1,353 W/m^2, or 1.8 $horsepower/m^2$.

Design an experiment to measure how much solar power reaches your part of the earth at any given time. If you keep records for a few years, you may notice some variation. Scientists have found as much as a 3% variation that depends on sunspots. Disturbingly, there is as much as a 50% variation in the amount of dangerous ultraviolet light that reaches the surface of the earth at the same time.

Finally, for the truly latest word on solar energy, try the internet site http://www.solstice.crest.org/ This site contains great tips, links to other sites, and practical information. A good store for purchasing voltaic cells and other solar equipment and books is Real Goods. This company has stores in many cities and also sells its products by mail order.

CHAPTER 7

SAVING ENERGY AT HOME AND SCHOOL

By reducing our use of energy, we extend the lifetime of conventional energy sources such as fossil fuels, improve the quality of the air we breathe, and reduce this nation's dependence on foreign energy sources as well as our energy bills.

In this chapter you'll have the opportunity to try some energy-saving projects and to discover some methods that could help your family and school reduce their use of energy and, therefore, the money they spend on it.

PROJECT 34: ELECTRICAL APPLIANCES AND ENERGY CONSERVATION

There are many electrical appliances in your home and school. To find the energy needed to operate an appliance for one year, you need to know two things: its power rating in watts and the number of hours the appliance is used.

Usually, you can find the wattage rating printed on the appliance itself or in the manual that came with it. In some cases you may find a plate that gives the current and voltage at which the appliance operates. Given this information, how could you find the power?

Next you will have to make reasonable estimates of the time that each appliance is used during a year. From your estimates of time and the power ratings, how can you determine the energy, in kilowatt-hours, that each appliance requires per year?

Make a survey of all the appliances in your home or school. How much energy is required to operate each one? Based on the cost of electricity, which you can find on a recent bill from your power company, how much does it cost to operate each appliance for one year? How do your values compare with those in Table 6, which gives the average values for wattage and hours used for various appliances? The last column gives the operating cost per year based on a figure of 10 cents per kilowatt-hour. As more and more areas allow people to choose their power provider, find out competing rates and add extra columns to this chart for each supplier.

Suggest ways to help your family or classmates and teachers reduce the use of these appliances. For example, turn off lights when they are not needed, use a clothesline instead of a dryer to dry your laundry when the weather is clear, use a dishwasher only when it is full of dishes, turn off the water heater when your family or school goes on vacation, etc. Which makes more of a difference in energy usage, turning off lights or taking shorter showers? Does your family use towels once and then wash them? How much could be saved by adopting slightly different habits? How much money will be saved if your family follows all your suggestions? As a project, try expanding this list by checking the wattage ratings on

TABLE 6:
The power and energy usage of various appliances

Appliance	Average wattage	Average # hours used per year	KWh per year	$ per year
Kitchen				
Blender	390	40	15.6	1.56
Broiler	1,490	70	98	9.80
Coffee maker	900	120	108	10.80
Dishwasher	1,200	300	360	36.00
Range	12,200	100	1,220	122.00
Microwave	1,450	130	189	18.90
Toaster	1,200	35	42	4.20
Waffle iron	1,100	20	22	2.20
Refrigerator (12 cu ft.)	240	3,000	720	72.00
Frostless refrigerator (12 cu ft.)	320	3,800	1,220	121.60
Freezer (12 cu ft.)	340	3,500	1,190	119.00
Frostless freezer (15 cu ft.)	440	4,000	1,760	176.00
Laundry				
Clothes dryer	4,800	200	960	96.00
Iron	1,000	140	140	14.00
Washing machine	500	200	100	10.00
Water heater	2,500	1,600	4,000	400.00
Quick-recovery				
Water heater	4,500	1,000	4,500	450.00

Comfort				
Air conditioner	900	1,000	900	90.00
Electric blanket	180	830	149	14.90
Dehumidifier	250	1,500	375	37.50
Fan (attic)	370	800	296	29.60
Fan (window)	200	850	170	17.00
Humidifier	180	900	162	16.20
Health and Beauty				
Hair dryer	750	50	37.5	3.80
Shaver	14	80	1.12	0.12
Sun lamp	280	60	16.8	1.68
Toothbrush	7	60	0.42	0.04
Entertainment				
Radio	70	1,200	84	8.40
TV (B/W, tube)	160	2,200	350	35.20
TV (B/W, solid state)	55	2,200	121	12.10
TV (color, tube)	300	2,200	660	66.00
TV (color, solid state)	200	2,200	440	54.00
Housewares				
Clock	2	8,760	17.5	1.76
Vacuum cleaner	630	75	47	4.72
Sewing Machine	75	140	10.5	10.50
Lighting				
Lightbulbs (11 60-watt in home)	660	1,500	1,000	100.00

other appliances around your home and school. See if you can keep track of how long these are on each day. For a dramatic science project display, find out how much coal must be burned, how much crude oil burned, or how much water must run through a hydroelectric dam powerhouse to produce the electricity needed to power each item for a year. This whole set of calculations will be much easier if you use a spreadsheet.

If your family or school decides to follow your suggestions, you can compare the electrical energy used after your suggestions were put into effect with the energy used before. How can you determine how much money has been saved because of your project? Why is a comparison of this year's energy bill with last year's a poor way to estimate savings? Check with your local utility about the concept of "heating degree-days" and "cooling degree-days." What figures should be compared? Your science teacher can order a CD-ROM called "The Sun's Joules," a disk full of good graphics, lots of data, and instructional materials about the sun. The disk is available at http://solstice.crest.org/renewables/SJ/. Show the teacher the site and the two of you can work its information into a very good project.

PROJECT 35: SAVING HEAT

One way to reduce energy use in the winter is to turn back the thermostat(s) in your home or school to 18°C (65°F) during the daytime and to 10°C (50°F) at night or when people are not in the building. (If you have electric heat, this could be included as part of Project 36.)

To see what impact this has on your heating bill, determine the amount of gas, oil, or electricity that you used last year for heating with the amount used after you instituted a thermostat set-back program. Then divide total fuel used for each winter by the number of degree-days in that winter.

If you use gas or electricity to generate heat and for other jobs that require energy, compare winter electric or gas bills with summer bills. The difference is probably due to heating. Calculate the money saved because of thermostat set-back. Automatic thermostats can perform this function, giving you different temperatures at different times during the days and nights. They can be programmed for different weekend settings. These cost under $25 and can be installed in under 15 minutes by anyone with a screwdriver. The wires on a thermostat carry very small currents, about battery power. Indeed, these thermostats have AA battery backups. What is the "payback" time on such a device? How long would it take to pay for itself and then start saving money for you?

PROJECT 36: BE A HOUSE DOCTOR

What You Need	
Checklist	Charts
Thermometer	Clipboard (it looks more official that way)
Draftometer	Portable radio and stethoscope (optional)
Measuring tape	
Calculator	

Doctors give their patients physical examinations. Armed with the knowledge you gleaned from the projects in this chapter and Chapter 3, you can be a house doctor, one who examines not people, but houses. A house doctor carries a checklist, a thermometer, a draftometer (a narrow strip of plastic wrap that moves in air currents and can be used to detect infiltration), a measuring tape, a pocket calculator, and some handy charts. After giving a house a thorough examination, the "doctor" may suggest to a homeowner ways to reduce energy use. Suggestions may involve reducing conductive heat losses, infiltration, and/or various conservation measures. Use the ideas from "The School Energy Doctor" as a guide. One trick for finding air leaks: play loud music inside a house and use a stethoscope around the windows on the outside. Hearing loud music will pinpoint air leaks.

As a house doctor you might begin your examination by using a series of energy checklists to be sure you check all the ways that energy may be wasted in the house you are examining. You might give a copy of your checklists to the owner. The owner will view these as suggestions that can be implemented if he or she chooses to do so. After your examination, point out to the owner any serious energy wasters that you have discovered.

Here are some checklists to use. You may want to add to, or delete from, these suggested checklists as you see fit.

Around-the-house checklist

- Is the dishwasher used only when filled? After the rinse cycle, are the dishes allowed to air-dry rather than using the electric heater in the washer? If dishes are washed by hand, is cold water or a pan of warm water used to rinse dishes rather than a stream of hot water?
- Is the refrigerator set at 40°F? A colder temperature is not necessary.
- Are the gaskets around the refrigerator and freezer doors tight enough to hold a dollar bill firmly?

An energy audit may turn up holes like this one through which heat can escape and cold air can enter.

- Is the freezer at 0°F? Is it full? (It's uneconomical to operate an empty or half-filled freezer. Put milk jugs of water in to fill the extra space.)

- Are clothes washers and dryers operated only for full loads? Alternatively, is the washer load-size setting used appropriately for smaller amounts of clothes?
- Are clothes washed only in cold or warm water when possible? Are they rinsed in cold water?
- Are soiled clothes soaked before washing?
- When large quantities of clothes need to be dried, they should be dried in consecutive loads so that the dryer's mass has to be heated only once.
- Is the lint screen cleaned after each load is dried?
- Is the outside vent cleaned frequently?
- Are clothes dried on a clothesline in clear weather?
- Are hot-water pipes insulated to reduce heat losses? Are cold-water pipes under the house also insulated to avoid freezing and to keep the water entering the water heater warmer?
- Are hot-water tanks wrapped in insulating blankets? There are kits available for this (but be careful—some tanks are already thoroughly insulated and it may be dangerous to add another layer of insulation. Check the warnings on the tanks).
- Is the water heater turned off when occupants are away for long periods?
- Is the thermostat setting on the water heater set higher than it needs to be?
- Do the occupants take quick showers rather than baths to reduce hot-water use?
- Are lights and other electrical appliances turned off when they are not in use?
- Are lightbulbs and fixtures clean?
- Are outside lights turned off when they are not needed?
- Has fluorescent lighting been substituted for the more costly incandescent lighting wherever possible?
- Do showers have flow reducers?

- Are there leaky faucets?
- Are doors to cooler areas kept closed?
- Are attic vents above ceiling insulation open? They should be. The air currents carry away moisture that diffuses into the attic.
- Are exhaust fans used only when necessary?
- Are windows kept closed when heating or air conditioning systems are on?

A heating and cooling checklist

- Are thermostats turned down at night and when there is no one at home? Are they turned to 78°F or above in the summer if there is central air conditioning? Is an automatic thermostat timer installed?
- If there is a fireplace, is the damper closed when the fireplace is not in operation?
- Is the fireplace used on cool and not on very cold days?
- Is the heating system equipped with an automatic damper to prevent hot air from escaping up the chimney when the burner is off?
- Is the heating system checked and adjusted at the beginning of every heating season?
- Are heating pipes and ducts that pass through unheated spaces insulated?
- Are the air filters on forced-air furnaces changed or cleaned during the heating season?
- Are radiators or baseboard heaters clean? Are they free of obstructions so that warm air can circulate, thus creating convection currents?
- Is water drained from a hot-water furnace periodically to remove rust and particles that might interfere with circulation?
- If hot-water radiators are present, is air "bled" from them periodically?

- Are air-conditioned or unheated rooms closed off from the others?

An insulation checklist

- Is the ceiling well insulated? Is the attic above it well ventilated?
- Are the outside walls insulated? (To find out, turn off the main electric switch, remove the cover plate from an electric outlet on an exterior wall, and look to see if there is insulation in the wall.)
- Are floors over cold areas insulated and carpeted?
- If there is a door to the attic, is it covered with insulation?

Insulation can keep a house or other building cooler in summer and warmer in winter.

An infiltration checklist

- Are there storm doors and windows?
- Are shades and draperies, preferably insulated ones, drawn at night and on cold, dark days? Do they fit tightly?
- Are shades and draperies drawn in summer to reduce solar heating?
- Do the eaves on the south side of the building shade the windows during the summer but not the winter?
- Are doors and windows caulked and weather-stripped? Do they fit tightly?
- Does your draftometer indicate infiltration around doors and windows or wall cracks?
- Are there entrance vestibules to reduce infiltration during entrance and exit?
- Are there windbreaks on the north and northwest sides of the house?

AN ENERGY AUDIT

In Chapter 3 you learned how to determine heat losses due to infiltration and conduction. You can use that knowledge to do an energy audit of the building you are examining.

From the volume of the building and the number of air turnovers per hour (N), which you can obtain by using Table 2 in Chapter 3, the heat required to change a cubic foot of air by 1°F (0.018 Btu/ft.^{3}/°F), and the number of degree-days (DD) in a heating season, you can determine the annual heat losses due to infiltration. You may remember that it can be found by the following formula:

$$\text{Heat losses (Btu)} = \text{Volume} \times 0.018 \times N \times 24 \times \text{DD}$$

To find the annual heat losses from a building due to

conduction, you will need to find the surface areas of ceilings, floors, and walls, and the R values of the materials that cover and lie within these surfaces. The R values of a variety of materials can be found in Appendix 1.

The air film on the external surface of a building has an R value of 0.17; the film on the interior surface has an R value of 0.68. Any enclosed air space surrounded by building materials has an R value of 0.9. If the space is bounded by aluminum foil facing inward, the R value becomes 2.17.

The R value for a wall with an internal cover of 3/8-in. gypsum board, filled with 3 1/2 inches of mineral wool insulation with aluminum foil facing, and covered externally by 3/4-in. wood sheathing, building paper, and 1/2-in. clapboards can be calculated as follows:

(air film) 0.68 + (gypsum board) 0.32 + (air space bounded by aluminum foil) 2.17 + (insulation) 11 + (sheathing) 1.0 + (building paper) 0.06 + (clapboards) 0.81 + (air film) 0.17 = total R value = 16.21 $ft.^2$–h–°F/Btu.

Before calculating heat losses through the wall area, you will have to subtract the total area of the windows and doors set in the walls.

The heat losses through the surfaces of doors and windows will have to be determined separately after you have established their R values.

A spreadsheet that you can design will help you organize your data and make calculations easier and more accurate.

POTENTIAL SAVINGS

Once you have determined the annual heat losses from the building, you can prescribe changes that will reduce them and thereby provide significant savings for the owner.

For example, if there is considerable infiltration due to poor caulking and weather stripping, the cost of materials to repair such problems is so insignificant that it is almost always worth the cost. Adding 7 in. of mineral wool insulation to an uninsulated ceiling might cost $300. If such a procedure reduces heat losses by 90 million Btu per heating season, then about 900 fewer gallons of fuel oil would be used. If the oil sells for a $1.50 a gallon, the owner would get his or her money back in one-fourth of a heating season—a deal that's hard to beat! Check with your local oil dealer to find out the efficiency rating on your furnace and the cost of oil delivered in your town. New "flame-retention" oil furnaces are so efficient it might take a lot longer to get payback. On the other hand, heating oil might become more expensive in the future.

If the ceiling already has 7 in. of insulation, another 7 inches might save only $50 per year. Would the owner be interested in doing this if the payback period is six years? Maybe, if he or she plans to stay in the house for many years. Maybe not, if the owner might move in a year or two. But only after an energy audit can a homeowner make rational decisions about spending money to reduce energy costs. In some areas, a local utility company will do an energy audit for free or at a very low cost. After you do your own audit, follow the utility company auditor around and see what he or she does and the results the auditor comes up with. Did you find more savings or less? Do you have a future in energy conservation?

GLOSSARY

absolute zero—0° Kelvin (–273°C); the temperature at which all molecular movement ceases.

active solar—energy generated by a photovoltaic cell, or usable energy for heating.

active solar energy system—a system designed to convert solar radiation into usable energy for space, water heating, or other uses. It requires a mechanical device, usually a pump or fan, to collect the sun's energy.

air film—a layer of still air adjacent to a surface that provides some thermal resistance.

alcohol fuels—fuels containing alcohol, such as gasahol or various alcohols used as fuels in their pure forms.

alternating current—(AC) flow of electricity that constantly changes direction between positive and negative sides. Almost all power produced by electric utilities in the United States moves in current that shifts direction at a rate of sixty times per second. In Europe, the rate is fifty times per second. One country, Iceland, uses only direct current. Almost all AC is converted to DC for actual use in most devices.

alternative fuels—(for vehicles) as defined by the National Energy Policy Act, the fuels are—methanol, denatured ethanol and other alcohols, separately or in mixtures of 85% by volume or more (or other percentage not less than 70% as determined by U.S. Department of Energy rule) with gasoline or other fuels; CNG; LNG; LPG; hydrogen; "coal-derived liquid fuels"; fuels "other than alcohols" derived from "biological materials;" electricity, or any other fuel determined to be "substantially not petroleum" and yielding "substantial energy security benefits and substantial environmental benefits."

ambient air temperature—surrounding temperature, such as the outdoor air temperature around a building.

ampere (Amp)—the unit of measure that tells how much electricity flows through a conductor. It is like using cubic feet per second to measure the flow of water. For example, a 1,200 watt, 120-volt hair dryer pulls 10 amperes of electric current (watts divided by volts).

ampere-second—a unit of electric charge equal to a coulomb. One ampere of electricity flowing for one second.

angle of incidence—the angle that the sun's rays make with a line perpendicular to a surface. The angle of incidence determines the percentage of direct sunshine intercepted by a surface.

animal waste conversion—process of obtaining energy from animal wastes. This is a type of biomass energy.

anthracite coal—coal with 90% carbon, very high heating value, and very few impurities.

barrel—in the petroleum industry, a barrel is 42 U.S. gallons. One barrel of oil has an energy content of 6 million British thermal units. It takes one barrel of oil to make enough gasoline to drive an average car from Los Angeles to San Francisco and back (at 18 miles per gallon over the 700-mile round trip).

bioconversion—processes that use plants or microorganisms to change one form of energy into another. For example, an experimental process uses algae to convert solar energy into gas that could be used for fuel.

biomass—organic material such as wood, grain, etc., that is a source of renewable energy. It also includes algae, sewage, and other organic substances that may be used to make energy through chemical processes.

British thermal unit (Btu)—energy required to raise the temperature of one pound of water one degree Fahrenheit. One Btu is equivalent to 252 calories, 778 foot-pounds, 1,055 joules, and 0.293 watt-hours.

calorie (cal)—the amount of heat needed to raise the temperature of one gram of water by one degree Celsius. One energy calorie is equivalent to 4.2 joules. Thus, it takes 500,000 calories of energy to boil a pot of coffee. One food Calorie equals 1,000 calories or 1.0 kilocalorie.

carbon dioxide (CO_2)—a colorless, odorless, nonpoisonous gas that is a normal part of the air. Carbon dioxide is exhaled by humans and animals and is absorbed by plant photosynthesis and by cold ocean water.

carbon monoxide (CO)—a colorless, odorless, highly poisonous gas made up of carbon and oxygen. It is formed by the incomplete combustion of carbon or carbonaceous material, including gasoline. It is a major air pollutant on the basis of weight.

caulking—material used to make an air-tight seal by filling in cracks, such as those around windows and doors.

Celsius (C)—a temperature scale based on the freezing (0 degrees) and boiling (100 degrees) points of water. Formerly known as Centigrade. To convert Celsius to Fahrenheit, multiply the number by 9, divide by 5, and add 32. For example—

10° Celsius × 9 = 90; 90 / 5 =18; 18 + 32 = 50° Fahrenheit.

chemical energy—energy stored in the chemical bonds of molecules.

circuit—one complete run of a set of electric conductors from a power source to various electrical devices (appliances, lights, etc.) and back to the same power source.

coal—a fossil fuel comprised primarily of carbon formed by the decomposition of plant matter in nonmarine environments millions of years ago.

coal gasification—process by which coal is converted into synthetic natural gas.

coal liquefaction—the process of converting coal into syncrude, or synthetic crude oil.

cogeneration—cogeneration uses the waste heat created by one process, for example during manufacturing, to produce steam that is used, in turn, to spin a turbine and generate electricity.

coke—a porous solid left over after the incomplete burning of coal or of crude oil. Almost pure carbon, with no impurities, it is used in making high-quality steel.

conduction—the transfer of heat energy through a material (solid, liquid, or gas) by the motion of adjacent atoms and molecules without gross displacement of the particles.

conductivity (thermal)—the ability of a substance to conduct heat. It is proportional to the cross-sectional area, the temperature difference, and the time, and inversely proportional to length.

conservation—steps taken to cause less energy to be used than would otherwise be the case. These steps may involve improved efficiency, avoidance of waste, reduced consumption, etc. They may involve installing equipment (such as a computer to ensure efficient energy use), modifying equipment (such as making a boiler more efficient), adding insulation, changing behavior patterns, etc.

convection—transferring heat by means of the motion of particles of liquid or gas.

cord—a measure of volume, 4 by 4 by 8 feet, used to define amounts of stacked wood available for use as fuel. Burned, a cord of wood produces about 5 million calories of energy.

cracking—a high-temperature process in which large molecules in petroleum (like tar) are cracked into smaller, more useful molecules like gasoline.

crude oil—the form in which oil is initially extracted, which is a mixture of hydrocarbons with some oxygen, nitrogen, and sulfur impurities; a fossil fuel.

cubic foot—the most common unit of measurement of natural gas volume. It equals the amount of gas required to fill a volume of one cubic foot under stated conditions of temperature, pressure, and water vapor. One cubic foot of natural gas has an energy content of approximately 1,000 Btus. One hundred (100) cubic feet equals one therm ($100 \text{ ft.}^3 = 1$ therm).

electrical energy—the energy associated with movement of electrons through a wire or circuit.

electric generator—a device that converts heat, chemical, or mechanical energy into electricity.

electric radiant heating—a heating system in which electric resistance is used to produce heat, which radiates to nearby surfaces. There is no fan component to a radiant heating system.

electricity—a property of the basic particles of matter. A form of energy having magnetic, radiant, and chemical effects. Electric current is created by a flow of charged particles (electrons).

electromagnetic radiation—radiation that is emitted in the form of photons, i.e., light, radio waves, ultraviolet, radar, infrared, x rays, cosmic rays.

endothermic—a reaction that takes heat in from the environment, that is, heat is absorbed by the system.

energy—the ability to do work. Forms of energy include thermal, mechanical, electrical and chemical. Energy may be transformed from one form into another.

ethanol (also known as ethyl alcohol or grain alcohol, CH_3CH_2OH)—a liquid that is produced chemically from ethylene or biologically from the fermentation of various sugars from carbohydrates found in agricultural crops and cellulosic residues from crops or wood. Used in the United States as a gasoline octane enhancer and oxygenate (for pollution control), it increases octane rating by 2.5 to 3.0 numbers at 10 percent concentration. Ethanol can also be used in higher concentration in vehicles optimized for its use by a change in carbueration.

evaporative cooling—cooling by exchange of latent heat from water sprays, jets of water, or wetted material.

exothermic—a chemical reaction that gives off heat to the environment, that is, heat is released from the system.

Fahrenheit—a temperature scale in which the boiling point of water is 212 degrees and its freezing point is 32 degrees. To convert Fahrenheit to Celsius, subtract 32, multiply by 5, and divide the product by 9. For example—100° Fahrenheit–32 = 68; 68 × 5 = 340; 340 / 9 = 37.8° C.

first law of thermodynamics—the total amount of energy and mass in the universe is constant; energy and mass can be neither created nor destroyed, although, under some conditions (nuclear), they can be converted from one to the other.

fission—a release of energy caused by the splitting of an atom's nucleus. This is the energy process used in conventional nuclear power plants to make the heat needed to run steam electric turbines.

fossil fuels—general term referring to fuels that have been generated by "fossilized" plant and animal matter over millions of years, i.e., coal, oil, and natural gas.

fractional distillation—method by which crude petroleum is refined into usable products on the basis of their different boiling points.

frequency—the number of cycles per second. Standard electric utility frequency in the United States is 60 cycles per second, or 60 Hertz.

fusion energy—a power source, now under development, based on the release of energy that occurs when atomic nuclei combine under the most extreme heat and pressure to form heavier nuclei. It is the energy process that fuels the sun and the stars.

gasohol—fuel made by distilling grain, wood, or other plant products into ethyl alcohol and mixing the alcohol with gasoline.

generator—a device consisting of a magnet and a coil of wire that changes mechanical energy into electrical energy.

geothermal energy—energy from the inner core of the earth; specifically from hot, molten rock pushing through the crust near the surface of the earth that heats underground water.

gigawatt (GW)—one billion watts (1,000,000,000 watts) of electric power. One gigawatt is enough to supply the electric demand of about one million average homes.

greenhouse effect—phenomenon in which oxides of nitrogen and carbon trap the energy radiated from the earth.

greenhouse gases—oxides of nitrogen, sulfur, and carbon as well as chlorofluorocarbon (CFC) compounds that absorb infrared radiation from the earth, causing global warming. For earth, the greenhouse gas with the greatest effect is water.

head—term used to describe the height of falling water in a hydropower system.

heat—energy transferred as a result of temperature differences. This results in more rapid movement of molecules or atoms.

heat capacity—the amount of heat necessary to raise the temperature of a given mass one degree. Heat capacity may be calculated by multiplying the mass by the specific heat.

heat loss—a decrease in the amount of heat contained in a space resulting from heat flow through walls, windows, roof, and other building surfaces, and from exfiltration of warm air.

heat pump—an air-conditioning unit that is capable of heating by refrigeration, transferring heat from one (often cooler) medium to another (often warmer) medium, and which may or may not include a capability for cooling. This reverse-cycle air conditioner usually provides cooling in summer.

heat transfer—flow of heat energy induced by a temperature difference. Heat typically flows from a heated or hot area to a cooled or cold area.

heating degree-day—a unit that measures the space heating needs of a location during a given period of time.

hertz—a unit of frequency equal to one cycle per second. Named after Heinrich R. Hertz.

horsepower (hp)—a unit for measuring the rate of doing work. One horsepower equals about three-fourths of a kilowatt (745.7 watts).

hydropower—energy from flowing water used for mechanical purposes or for electricity production.

infiltration—the uncontrolled inward leakage of air through cracks and gaps in the building envelope, especially around windows, doors, and duct systems.

insolation—the total amount of solar radiation (direct, diffuse, and reflected) striking a surface exposed to the sky.

insulation—process in which a material slows heat loss or gain; also, a substance that insulates.

joule—one Newton-meter; a unit of work equivalent to 0.239 calories.

kilovolt (kv)—one thousand volts (1,000). Electrical distribution lines in residential areas usually are 12 kv (12,000 volts).

kilowatt (kW)—one thousand (1,000) watts. A unit of measure of the amount of electricity needed to operate given equipment. On a hot summer afternoon, a typical home, with central air conditioning and other equipment in use might have a demand of four kW each hour.

kilowatt-hour (KWh)—the most commonly used unit for measuring the amount of electricity consumed. It means one kilowatt of electrical power supplied for one hour. In 1989, a typical California household consumed 534 kWh in an average month.

kinetic energy—energy of motion, $\frac{1}{2} mv^2$.

latent heat—a change in heat content that occurs without a corresponding change in temperature, usually accompanied by a change of state (as from liquid to vapor during evaporation).

mechanical energy—energy that can be used directly to do work, either potential or kinetic.

megawatt (MW)—one thousand kilowatts (1,000 kW). One megawatt is enough energy to power 1,000 average homes.

methane (CH_4)—the simplest of hydrocarbons and the principal constituent of natural gas. Pure methane has a heating value of 1,012 Btu per standard cubic foot.

methanol (also known as methyl alcohol, wood alcohol, CH_3OH)—a liquid formed by catalytically combining carbon monoxide (CO) with hydrogen (H_2) in a 1:2 ratio by volume under high temperature and pressure. Commercially, it is typically made by steam-reforming natural gas. Also formed in the destructive distillation of wood, hence one of its names.

microwave—electromagnetic radiation with wavelengths of a few centimeters. It falls between infrared and radio wavelengths on the electromagnetic spectrum. One particular frequency makes water molecules spin, creating frictional heat, as in microwave ovens.

natural gas—methane with about 1% other light hydrocarbons; a fossil fuel. It is the product of the anaerobic decomposition of organic matter, intestinal fermentation in animals (in humans, especially after a meal of beans), and is one of the greenhouse gases.

newton—a unit of force. The amount of force it takes to accelerate one kilogram at one meter per second per second.

NO_x—oxides of nitrogen that are a chief component of air pollution that can be produced by the burning of fossil fuels. Also called nitrogen oxides.

nuclear energy—the energy stored in the nucleus of an atom that can be released upon fission or fusion. Currently, we can create "controlled fission" in power plants but only "uncontrolled" fusion (in bombs).

ocean thermal gradient (OTG)—temperature differences between deep and surface water. Deep water is likely to be 25 to 45° F colder. The term also refers to experimental technology that could use the temperature differences as a means of producing energy.

ohm (Ω)—a unit of measure of electrical resistance. One volt can produce a current of one ampere through a resistance of one ohm.

OPEC—Organization of Petroleum Exporting Countries, founded in 1960 to unify and coordinate petroleum polices of the members. Headquarters is in Vienna, Austria.

ozone—a kind of oxygen that has three atoms per molecule instead of the usual two (O_3 instead of O_2). Ozone is a poisonous gas, but the ozone layer in the upper atmosphere shields life on earth from deadly ultraviolet radiation from space. The UV light creates the ozone layer by reacting with oxygen molecules. Ozone is also produced by sunlight shining on certain types of smog, by electrical sparks, and lightning bolts.

particulate matter—unburned fuel particles that form smoke or soot and stick to lung tissue when inhaled. A chief component of exhaust emissions from diesel engines and woodstoves.

passive solar heating—using a material to collect and store thermal energy from the sun by means of architectural design (such as arrangement of windows) and materials (such as floors that store heat, or other thermal mass).

photon—a massless particle of electromagnetic energy (light).

photosynthesis—the production of glucose in a plant from water and carbon dioxide using solar radiation.

photovoltaic cell—a semiconductor that converts light directly into electricity.

potential energy—stored energy in a system that is a function of position or chemical bonds.

power—energy per unit of time. One joule of energy used to do work over a period of one second is called a watt.

QUAD—an amount of energy equal to one quadrillion (1,000,000,000,000,000) British thermal units (Btus). An amount of energy equal to 170 million barrels of oil. Total U.S. consumption of all forms of energy in the 1990s was about 83 quads in an average year.

R-value—a unit of thermal resistance used for comparing insulating values of different material. It is basically a measure of the effectiveness of insulation in stopping heat flow. The higher the R-value number for a material, the greater its insulating properties and the slower the heat flow through it.

radiant energy—energy transferred by the exchange of electromagnetic waves from a hot or warm object to one that is cold or cooler. Direct contact with the object is not necessary for the heat transfer to occur, as is seen in energy coming to earth from the sun.

renewable resource—an energy resource in the environment which can be renewed if proper care is taken. Examples include hydropower, wind power, biomass, solar power, and geothermal energy.

reserve—the amount of a resource that is recoverable.

resistance (electrical)—the ability of all conductors of electricity to resist the flow of current, turning some of it into heat. Resistance depends on the cross section of the conductor (the smaller the cross section, the greater the resistance) and its temperature (the hotter the cross section, the greater its resistance).

second law of thermodynamics—overall, the disorder in the universe always increases and objects at a high energy state go to a lower energy state.

short ton—2,000 lb. A long ton is 2,200 lb. and is actually a metric tonne. A short ton of coal can provide about 26×10^6 Btu.

smog—smoky fog that hangs in the atmosphere as a result of the burning of fossil fuels with impurities.

solar cell—a photovoltaic cell that can convert light directly into electricity. A typical solar cell uses semiconductors made from silicon.

solar collector—a surface or device that absorbs solar heat and transfers it to a fluid. The heated fluid is then used to move the heat energy to where it will be useful, such as in water or space-heating equipment.

solar energy—energy radiated from the sun, primarily heat and light, and used as energy on earth.

specific heat—in English units, the quantity of heat, in Btu, needed to raise the temperature of 1 lb. of material 1° F. The metric units are calories per degree per gram (cal/°-g), or joules per degree per gram (j/°-g).

temperature—the average speed of all the molecules within a certain volume. The degree of hotness or coldness is usually measured on one of several arbitrary scales based on some observable phenomenon (such as expansion) in degrees Fahrenheit, Celsius, or Kelvin.

THERM—a measure of heat energy equal to 100,000 Btu.

thermal energy—energy in the form of heat.

thermal mass—a heat storage material, such as water or masonry, used in passive solar heating systems, which radiates heat to the surroundings after the sun goes down.

volt—a unit of electromotive force. It is the amount of force required to drive a steady current of one ampere through a resistance of one ohm. Particularly, it is equal to 1 joule of energy per coulomb of charge. Electrical systems in most homes have 120 volts in the United States, and 240 volts in Europe.

watt—a unit of measure of electric power at a point in time, as capacity or demand. It is one joule of energy used over one second. The Watt is named after Scottish inventor James Watt (who improved the steam engine) and is capitalized when shortened to W and used with other abbreviations, as in kWh.

watt-hour—one watt of power expended for one hour. On home electricity bills, you usually find kilowatt-hours—a thousand watt-hours.

wind power—energy per time from the moving air used to turn large windmills that generate electricity.

work—a force applied to an object over a certain distance. Work = Force × Distance. The unit of work in the English system is the foot-pound; in the metric it is the joule.

APPENDIX 1

R Values of Various Materials

Material	Thickness (inches)	R Value (sq ft.-h-°F/Btu)
Insulation		
mineral wool batts	1	3.1
mineral wool batts	3.5	11
mineral wool batts	6	19
mineral wool batts	12	38
glass fiber	1	2.2
rock wool	1	2.7
cellulose (paper)	1	3.7
vermiculite	1	2.2
perlite	1	2.7
expanded polyurethane	1	5.9
expanded polystyrene	1	4.7
polyisocyanurate sheathing	1	8.0
Building Materials		
wood sheathing	0.75	1.0
plywood	0.50	0.63
bevel-lapped siding	0.50	0.81
gypsum board	0.375	0.32
building paper	—	0.06

vapor barrier (plastic)	—	0
wood shingles	—	0.87
asphalt shingles	—	0.44
linoleum	—	0.08
carpet with fiber pad	—	2.1
hardwood floor	—	0.71
Windows and Doors		
single-glazed window	—	1.0
double-glazed window	—	2.0
exterior door (wood)	—	2.0
Masonry		
concrete block	8.0	1.1
brick (common)	4.0	2.0
concrete (poured)	8.0	0.64
Air Films and Spaces		
air space, between building materials	0.75 or more	0.90
air space, between aluminum foils	0.75 or more	2.17
air film, on exterior surface	—	0.17
air film, on interior surface	—	0.68

APPENDIX 2

Specific Heats of Some Common Substances

Specific Heat Material	(cal/g/°C)
Air	0.24
Aluminum	0.214
Asbestos wool	0.20
Asphalt	0.22
Brick	0.20
Cast iron	0.12
Clay	0.22
Concrete	0.22
Copper	0.092
Glass	0.18
Glass wool	0.157
Gypsum	0.26
Ice (0°C)	0.487
Limestone	0.217
Marble	0.21
Sand	0.191
Steel	0.12
Stones	0.21
Water	1.00
White fir	0.65
White oak	0.57
White pine	0.67

APPENDIX 3

Resources

EQUIPMENT

Many of the experiments in this book recommend the use of electronic probes that can be hooked up to a computer or some calculators. These allow for precise measurements that can be very difficult to accomplish any other way. They also will automatically record your data. The best source the authors know of for these is Vernier Software, 8565 S.W. Beaverton-Hillsdale Hwy., Portland, OR 97225-2429 (503-297-5317 or http://www.vernier.com, email: info@vernier.com). The prices are the most reasonable; you can construct some of the probes yourself; the company is small enough to answer questions; the probes work on Macs, PCs, Texas Instruments calculators, and a number of them even work on the old Apple //e computers your school may have in a storeroom some place. *Graphical Analysis* is one of their programs for handling data from experiments. It graphs your work as you enter the numbers and you can try "curve-fitting" to get a mathematical equation that will match your data. Science fair judges are really impressed by equations that match your data. You can get a site license for the whole school for this program for under $50. Talk to your science teacher about this.

BOOKS

Note: (OOP) means the book is Out of Print—check libraries or used book stores.

Adams, Richard C., and Robert Gardner. *Ideas for Science Projects*. Danbury, CT: Franklin Watts, 1997.

Adams, Richard C., and Robert Gardner. *More Ideas for Science Projects*. Danbury, CT: Franklin Watts, 1998.

Adams, Richard C., and Peter S. Goodwin. *Engineering Projects for Young Scientists*. Danbury, CT: Franklin Watts, 2001.

Adams, Richard C., and Peter S. Goodwin. *Physics Projects for Young Scientists*. Danbury, CT: Franklin Watts, 2000.

Alternative Energy Handbook. Emmaus, Pa.: Rodale Press, 1979.

Anderson, Bruce and Charles A. Banks. *Solar Building Architecture*. Cambridge, MA: M.I.T. Press, 1990.

Anderson, Bruce, with Michael Riodan. *The Solar Home Book*. Harrisville, NH: Brick House Publishing Company, 1976. (OOP)

Appel, Kenneth, John Gastineau, Clarence Bakken, and David Vernier. *Physics with Computers*. Portland, Ore.: Vernier Software, 1998. Very detailed instructions for using computer and calculator interfaces in 34 experiments.

Berry, R. Stephen. *Understanding Energy: Energy, Entropy, and Thermodynamics for EveryMan*. NY: World Scientific Publishing Co., 1991.

Burton, James, and Kim Taylor. *The Nature and Science of Energy*. NY: Gareth Stevens, 1998.

Davidson, Joel. *The New Solar Electric Home: The Photovoltaics How-To Book*. NY: Aztec Publications, 1987.

Eyewitness Books, Source: Energy. NY: D.K. Publishing, 1993. Great source for good pictures in your report.

Gardner, Robert. *Ideas for Science Projects*. NY: Franklin Watts, 1986.

———. *Save That Energy*. NY: Julian Messner, 1981.

Gipe, Paul. *Wind Power for Home and Business: Renewable Energy for the 1990's and Beyond*. NY: Chelsea Green Publishing Co., 1993.

Haber-Schaim, Uri. *Energy: A Sequel to IPS*. Englewood Cliffs, NJ: Prentice-Hall, 1977. (OOP)

In the Bank or Up the Chimney. Washington D.C.: U.S. Government Printing Office, 1977.

Kachadorian, James. *The Passive Solar House*. NY: Chelsea Green Publishing Co., 1997.

Keyes, John. *Harnessing the Sun to Heat Your Home*. Dobbs Ferry, NY: Morgan & Morgan, 1974. (OOP)

Komp, Richard J. *Practical Voltaics: Electricity from Solar Cells*. NY: Aztec Publications, 1995.

Lori, William S. *Winning with Science*, Sarasota, FL: Loiry Publishing 1991.

McNinch, Sandra. *Sun Power; a Bibliography of U.S. Government Documents on Solar Energy*. NY: Greenwood Publishing Group, 1981.

Morrison, James W. *The Complete Energy-Saving Home Improvement Guide*. NY: Arco, 1978. (OOP)

NSTA. *Science Fairs and Projects: Grades 7–12*. 1840 Wilson Blvd., Arlington, VA, 22201, National Science Teachers Association, 1990. Suggestions involving students in science projects and guiding parents in their child's investigation.

Oei, Paul D., Eugene W. Sorenson, and Chih-Ming Chang. *Projects & Experiments in Energy*. NY: National Energy Foundation, 1982. (OOP)

Parker, Steve. *Energy (Science Works Series)*. Gareth Stevens Publications, 1997.

Schubert, Robert C., and L.D. Ryan. *Fundamentals of Solar Heating*. Englewood Cliffs, NJ: Prentice Hall, 1981.

Smith, Norman. How to do Successful Science Projects. New York: Julian Messner, 1990.

Solar Energy: A Bibliography. SB-009. Washington D.C.: U.S. Government Printing Office, 1979.

Telkes, Maria. "Storage of Solar Energy." *ASHRAE Transactions*, vol. 80B, 1974.

Tips for Energy Savers. Washington D.C.: Federal Energy Administration, 1977.

Tocci, Salvatore. *How to do a Science Fair Project*. Danbury, CT: Franklin Watts, 1997.

Vazquez, Laura. *Not Just Another Science Fair: A Handbook and More for Science Fair Organizers*. Glenview, IL: Goodyear Books, Scott Foresman, 1994.

Walker, Harry O. *Energy: Options and Issues*. CA: University of California, 1977. (OOP)

Webster, David. *How to do a Science Project*. NY: Franklin Watts, 1974. (OOP)

Wilson, Alex. *Consumer Guide to Home Energy Savings*. NY: John Morrill, 1996.

Wolfe, Ralph, and Peter Clegg. *Home Energy for the Eighties*. Charlotte, VT.: Garden Way Publishing, 1979. (OOP)

Woodruff, John. *Energy Science Projects*. Raintree/Steck Vaughn, 1998.

INTERNET RESOURCES

Note: Internet sites change over time. If you get an error, try the address again. If it still doesn't work, start chopping off the right end of the address.

Advanced Communications. "ThinkQuest" http://www.thinkquest.org for students ages 12-19; http://www.advanced.org/thinkquestjr.html for grades 4-6. "ThinkQuest" science fair prizes are substantial, in the $5000 range for students, $10,000 for sponsoring teachers, and $5,000 for schools. Project must include Internet collaboration, however.

Alternative Energy, Engineering Design Guide http://www.alt-energy.com/aee

Alternative Fuels Data Center. Good government site for hard data. http://www.afdc.doe.gov

Amazon.Com. http://www.amazon.com. Great source for new books and is of some help in finding out-of-print books (see Powell's City of Books below).

American Association of Physics Teachers. "Publication Listing" http://www.unl.edu/physics/Education/toolkit/publicate.html. Good source for articles, books, and pamphlets.

American Solar Energy Society. http://www.ases.org/solar

Barron, J. "Science Fair Ideas." http://www.stemnet.nf.ca/~jbarron/scifair.html. Great place for ideas!

College of Agricultural and Environmental Sciences http://www.uga.edu/oasp/gsef/guide.html. Guidelines on how to conduct a research project for the young scientist.

Energy Efficiency and Renewable Energy Network (EREN). http://www.eren.doe.gov. This site has 917 pdf format documents on materials for saving energy. Go from the main site to "consumer info/energy-savers."

Engineering Competitions. "National Engineers Week" http://www.eweek.org

DETAILS ON COMPETITIONS AND PROJECTS

General Wolfe School. "Science Projects" http://www.wsd1.org/nnl/general_w/scdept/hotspots.htm. Good examples of projects.

Intel Corp. "Intel International Science and Engineering Fair" http://www.sciserv.org/isef. Intel ISEF is the Olympics of the science competitions. Thousands of students from all 50 states and 40 nations come together each May to compete for scholarships, tuition grants, internships, scientific field trips and the grand prize: a trip to attend the Nobel Prize Ceremonies in Stockholm, Sweden.

Mr. Solar Home Page. http://www.mrsolar.com/

National Science Teachers Association. http://www.nsta.org. Great source for books, pamphlets, and links to other sites.

Powell's City of Books, Portland, Oregon. Largest physical bookstore in the United States. Their on-line service: http://www.powells.com, lists all their used books also, unlike Amazon.com, and you can find many out-of-print books this way. If you get to Portland, visit this place. It takes up a city block, they issue maps at the entrance, and just their technical bookstore has to be put on another half-block. Take good walking shoes and money with you.

Real Goods. These stores are around the country and have all sorts of equipment and ideas for making your home energy independent. Good source for solar cells, ideas, and clever little goodies that can enhance your science project. http://www.realgoods.com

Solstice. One of the best sites for finding links and data about solar energy. http://solstice.crest.org/

State by State Energy Info. Use this site to get the data from your own state, what programs are operating, and compare with other states for your project rationale and write-up. http://204.243.73.5/STATES.htm

"The Source" for Renewable Energy, a comprehensive list of businesses by geographical area and products. http://www.rmii.com/the Source/renewableEnergy/

Ultimate Science Fair Resource, http://www.neltec.com/scifair/index.html

UMDNJ and Coriell Research Library. "Science Fair Projects: a Resource for Students and Teacher." http://arginine.umdnj.edu/~swartz/scifair.html. Good source for project ideas.

United Learning Co. How to Prepare a Science Fair Project 6633 W. Howard St., Niles IL 60648: United Learning, 1991. Provides teacher and student guides, which include sample forms, sample long-range schedule, ideas for science fair projects, judges score sheet, and certificate of accomplishment. Also includes handy videotape. http://www.unitedlearning.com/features/

University of Oregon Alternate Energy Course Information. This university offers on-line course materials. You can take the course or just use the materials as sources for your projects. http://zebu.uoregon.edu/phys162.html

University of Southern California. "Science Fairs." http://physics.usc.edu/~gould/ScienceFairs/. Virtual Library (USC) listing of every science fair they could find.

VITA Publication List. These are good pamphlets on simple energy projects for Third World countries. Linking your project to an international concern will impress judges. gopher://gopher.vita.org:70/11/intl/vita/pubcat/refer/energy

Your teacher or an adult who may have helped you with your project may be able to provide information about state and local science fairs if you are interested in entering one. Listed below are some of the national competitions related to science fairs or research. If you are interested, write to the address provided for more information.

Earthwatch offers research internships in various science projects throughout the world. Write to: Earthwatch, 680 Mt. Auburn St., P.O. Box 9104, Watertown, MA 02272. (800) 776-0188, (617) 926-8532 FAX. email: info@earthwatch.org. Online: http://gaia.earthwatch.org/

Intel Science Talent Search (formerly "Westinghouse Science Talent Search") provides scholarship awards to winners of this program, which involves independent research in the physical, biological, behavioral, and social sciences, as well as mathematics and engineering. Write to: Science Service, 1719 N Street NW, Washington, D.C. 20036. Online at http://www.sciserv.org/stshome.htm

International Science and Engineering Fair is sponsored by General Motors and nearly fifty other organizations. Projects may be any of twelve categories ranging from computers to zoology. Write to: Science Service, 1719 N Street NW, Washington, D.C. 20036. For year 2000 competition: http://www.isef2000.org/

Junior Engineering Technical Society holds a contest for projects in engineering. Write to: JETS, Inc., 1420 King St., Suite 405, Alexandria, VA 22314-2794. (703) 548-5387, (703) 548-0769 FAX. e-mail jets@nae.edu. On-line: http://www.asee.org/jets/

National Energy Foundation runs energy education programs and science fairs. Write to: National Energy Foundation, 5225 Wiley Post Way, Suite 170, Salt Lake City, UT 84116. (801) 539-1406, (801) 539-1451 FAX. On-line: http://www.nef1.org/

National Science Foundation Science Training Project is a program sponsored by NSF that offers science training at a variety of schools, colleges, and laboratories during the summer months. Write to: National Science Foundation, 4201 Wilson Blvd., Arlington, VA 22230. (703) 306-1234. On-line: http://www.nsf.gov/EHR/ESIE?index.html

INDEX

Italicized page numbers refer to illustrations.